AGRICULTURE ISSUES AND POLICIES

ORGANIC FARMING AND PEANUT CROPS

AGRICULTURE ISSUES AND POLICIES

Assessing Disease Potential in U.S. Aquaculture Industry
Emilija Kalnins (Editor)
2009. ISBN: 978-1-60741-543-5

Agricultural Conservation
Anthony G. Hargis (Editor)
2009. ISBN: 978-1-60692-273-6
2009. ISBN: 978-1-60876-813-4 (E-book)

Environmental Services and Agriculture
Karl T. Poston (Editor)
2009. ISBN: 978-1-60741-053-9

Weeds: Management, Economic Impacts and Biology
Rudolph V. Kingely (Editor)
2009. ISBN 978-1-60741-010-2

No-Till Farming: Effects on Soil, Pros and Cons and Potential
Earl T. Nardali (Editor)
2009. ISBN: 978-1-60741-402-5

Inspection and Protection of U.S. Meat and Poultry
Finn J. Amundson (Editor)
2009. ISBN: 978-1-60741-120-8

Global Beef Trade
Alessandro Ferrara (Editor)
2009. ISBN: 978-1-60741-121-5

Organic Food - Economics and Issues
Earl D. Straub (Editor)
2009. ISBN: 978-1-60741-130-7

Soybean and Wheat Crops: Growth, Fertilization and Yield
Samuel Davies and George Evans
2009. ISBN: 978-1-60741-173-4
2009. ISBN: 978-1-61668-984-1 (E-book)

Effects of Liberalizing World Agricultural Trade
Henrik J. Ehrstrom (Editor)
2009. ISBN: 978-1-60741-198-7
2009. ISBN: 978-1-60876-601-7 (E-book)

Hired Farmworkers: Profile and Labor Issues
Rea S. Berube (Editor)
2009. ISBN 978-1-60741-232-8

Agricultural Wastes
Geoffrey S. Ashworth and Pablo Azevedo (Editors)
2009. ISBN: 978-1-60741-305-9

Sugar Beet Crops: Growth, Fertilization & Yield
Claus T. Hertsburg (Editor)
2009. ISBN: 978-1-60741-491-9

Economic Impacts of Foreign-Source Animal Disease
Jace R. Corder (Editor)
2009. ISBN: 978-1-60741-601-2
2009. ISBN: 978-1-60876-602-4 (E-book)

Essential Oils: Art, Agriculture, Science, Industry and Entrepreneurship(A Focus on the Asia-Pacific Region)
Murray Hunter
2009. ISBN: 978-1-60741-865-8

Drivers and Restraints for Economically Efficient Farm Production
Helena Hansson Karin Larsén Bo Öhlmér
2009. ISBN: 978-1-60876-171-5

Corn Crop Production: Growth, Fertilization and Yield
Arn T. Danforth (Editor)
2009. ISBN: 978-1-60741-955-6
2009. ISBN: 978-1-60876-860-8 (E-book)

Phosphate Solubilizing Microbes for Crop Improvement
Mohammad Saghir Khan and Almas Zaidi (Editors)
2009. ISBN: 978-1-60876-112-8

Agriculture and Food
Claus Schäfer
2010. ISBN: 978-1-60692-038-1

Agricultural Economics: New Research
Tomas H. Lee (Editor)
2010. ISBN: 978-1-61668-077-0
2010. ISBN: 978-1-61668-445-7 (E-book)

U.S. Biobased Products Market Potential and Projections through 2025
Meredith A. Williamson (Editor)
2010 ISBN: 978-1-60741-033-1

Genetically Engineered Crops and Proposed Oversight
Brandon C. Vaughn (Editor)
2010 ISBN: 978-1-60876-067-1
2010 ISBN: 978-1-61728-094-8 (E-book)

The Sugar Industry and Cotton Crops
Peter T. Jenkins (Editor)
2010. ISBN: 978-1-61668-320-7
2010. ISBN: 978-1-61668-439-6 (E-book)

Tomatoes: Agricultural Procedures, Pathogen Interactions and Health Effects
Eric D. Aubé and Frederick H. Poole (Editors)
2010. ISBN: 978-1-60876-869-1

Sustainable Agriculture: Technology, Planning and Management
Augusto Salazar and Ismael Rios (Editors)
2010. ISBN: 978-1-60876-269-9

Organic Farming and Peanut Crops
Darren C. Grossman and Terrance L. Barrios (Editors)
2010. ISBN: 978-1-60876-187-6

Farm Bill of 2008: Major Provisions and Legislative Action
Mary T. Conner (Editor)
2010. ISBN: 978-1-60741-750-7

America's Family Farms
Efren J. Tamayo (Editor)
2010. ISBN: 978-1-60741-751-4

Million Dollar Farms in the New Century
Samuel D. Bosworth (Editor)
2010. ISBN: 978-1-60741-755-2
2010. ISBN: 978-1-61668-541-6 (E-book)

Pesticide Resistance, Population Dynamics and Invasive Species Management
Gregory J. McKee, Colln A. Carter, James A. Chalfant, Rachael E. Goodhue, and Frank G. Zalom
2010. ISBN: 978-1-60741-758-3

Agroforestry Development on the Canadian Prairies
Suren N. Kulshreshtha, Ken Van Rees, Hayley Hesseln, Mark Johnston, John Kort
2010. ISBN: 978-1-61668-266-8
2010. ISBN: 978-1-61668-588-1 (E-book)

Agriculture Research and Technology
Kristian Bundgaard and Luke Isaksen (Editors)
2010 ISBN: 978-1-60741-850-4

Soil Phenols
A. Muscolo and M. Sidari (Editors)
2010. ISBN: 978-1-60876-264-4

Ecophysiology of Tropical Tree Crops
Fabio DeMatta (Editor)
2010.: ISBN 978-1-60876-392-4

Agricultural Production
Felix C. Wager (Editor)
2010.: ISBN 978-1-61668-695-6
2010.: ISBN 978-1-61728-050-4 (E-book)

Rethinking Structural Reform in Turkish Agriculture: Beyond the World Bank's Strategy
Baris Karapinar, Fikret Adaman and Gokhan Ozertan (Editors)
2010. ISBN: 978-1-60876-718-2

Desalination: Methods, Costs and Technology
Irena A. Urbonienė (Editor)
2010. ISBN: 978-1-61668-909-4
2010. ISBN: 978-1-61728-242-3 (E-book)

Manure Use for Fertilizer and Energy
Connor D. Macias (Editor)
2010. ISBN: 978-1-60876-847-9

Price Dynamics behind Consumer Food Purchases
Morgan D. Fitzpatrick (Editor)
2010. ISBN: 978-1-60876-892-9

Governance of Agrarian Sustainability
Hrabrin Bachev
2010. ISBN: 978-1-60876-888-2

Transformation of U.S. Animal Agriculture
Justin M. Daigle (Editor)
2010. ISBN: 978-1-60876-938-4

The Peanut Plant and Light: Spermidines from Peanut Flowers
Victor S. Sobolev, James B. Gloe and Arlene A. Sy
2010. ISBN: 978-1-61668-028-2
2010. ISBN: 978-1-61668-371-9 (E-book)

Agriculture and Environmental Security in Southern Ontario's Watersheds
Glen Filson, Bamidele Adekunle, and Katia Marzall
2010. ISBN: 978-1-61668-156-2
2010. ISBN: 978-1-61668-372-6 (E-book)

Alignment-Free Models in Plant Genomics: Theoretical, Experimental, and Legal issues
Humberto González-Díaz, Guillermin Agüero-Chapin, Cristian Robert Munteanu, Francisco Prado-Prado, Kuo-Chen Chou, Aliuska Duardo-Sanchez, Grace Patlewicz and Antonio López-Diaz
2010. ISBN: 978-61668-333-7
2010. ISBN: 978-61668-603-1 (E-book)

Manure: Management, Uses and Environmental Impacts
Carmen S. Dellaguardia (Editor)
2010. ISBN: 978-1-61668-424-2
2010. ISBN: 978-1-61668-647-5 (E-book)

Postharvest Management of Litchi and Rambutan Fruits
D. Sivakumar and L. Korsten
2010. ISBN: 978-1-61668-688-8
2010. ISBN: 978-1-61728-089-4 (E-book)

Understanding U.S. Farm Exits, Farm Survival and the Role of Commodity Payments
Lucas A. Hamilton (Editor)
2010. ISBN: 978-1-61668-995-7
2010. ISBN: 978-1-61728-245-4 (E-book)

Agriculture Issues and Policies

Organic Farming and Peanut Crops

Darren C. Grossman
and
Terrance L. Barrios
Editors

Nova Science Publishers, Inc.
New York

For permission to use material from this book please contact us:
Telephone 631-231-7269; Fax 631-231-8175
Web Site: http://www.novapublishers.com

Library of Congress Cataloging-in-Publication Data

Organic farming and peanut crops / editors: Darren C. Grossman and Terrance L. Barrios.
p. cm.
Includes index.
ISBN 978-1-60876-187-6 (hardcover)
1. Peanuts--Organic farming. I. Grossman, Darren C. II. Barrios, Terrance L.
SB351.P3O69 2010
633.3'6884--dc22

2009048941

Published by Nova Science Publishers, Inc. ✝ New York

CONTENTS

PREFACE

Organic farming is the oldest way of farming, i.e., without the use of chemicals to enhance the production of meat, vegetables, milk and derivatives, and without the use of antibiotics and pesticides to decrease economical losses due to animal sickness and plant disease. Control of plant-parasitic nematodes is difficult, especially in organic farming systems, because in contrast to foliar diseases and insect pests, effective control methods and monitoring systems are not always available. Organic farmers struggle with nematode problems using cultural, physical and biological control methods. This book discusses effective management of plant-parasitic nematodes, which can be achieved by combining several control methods after identifying the nematodes in the field and understanding their biology. Moreover, in this book, the authors illustrate how participatory plant breeding (PPB) may answer the needs of organic agriculture, and how organic farmers and their organisations have met with researchers to build common experiments of plant breeding on farms. In addition, the early history and significance of the peanut crop is discussed. Annual world production of peanuts at 30 million tons makes this crop one of the most important agricultural commodities. The physiological and biochemical response of peanut microsimbionts under different types of stress are examined as well, and compared with rhizobia. This book also explores the concept of scientific paradigms and draws parallels with paradigms in agriculture, and in particular focuses upon organic farming and genetic engineering.

Chapter 1 - Control of plant-parasitic nematodes is difficult, especially in organic farming systems, because in contrast to foliar diseases and insect pests, effective control methods and monitoring systems are not always available. Organic farmers struggle with nematode problems using cultural, physical and biological control methods. Cultural practices such as crop rotations, use of resistant cultivars, and soil amendments, are very commonly integrated into the soilborne-disease control strategies in organic farming systems. Nematode-resistant varieties are available mainly for root-knot (*Meloidogyne* spp.) and cyst (*Heterodera* and *Globodera* spp.) nematodes in a few crops. Use of grafted vegetable seedlings has been become popular in Solanaceae and Cucurbitaceae crops, but no rootstocks resistant to root-knot nematodes are commercially available for the latter. Soil amendments, especially those with high nitrogen-to-carbon ratios, have been reported to exhibit nematicidal and fungicidal activity, mainly through the release of ammonia during their decomposition in the soil and/or through the consequent increased populations of antagonistic microorganisms. Nematode control efficacy of ammonia can be enhanced by manipulating soil environments, such as soil pH and temperature, and by suppressing microorganisms involved in nitrification. Plants that possess nematicidal compounds or their precursors are often incorporated to control soilborne diseases (biofumigation), with a special focus being placed on members of the Brassicaceae family. This effect is generally attributed to biocidal isothiocyanates, which are released from glucosinolates in plants by enzymatic activity. Other plants containing nematicidal compounds are useful as cover crops and/or green manure. The effects of composts on plant-parasitic nematodes have not been studied as thoroughly as their effects on fungal diseases. Soil solarization has been developed as a physical control method, and used to control soilborne diseases and weeds. This control method is also effective in nematode control; however, combinations of soil solarization and organic amendments greatly improve the nematode-control efficacy of soil solarization alone. Use of biological agents in nematode control is not as popular or successful as it is in insect control although many nematode-antagonistic organisms are known. Effective management of plant-parasitic nematodes can be achieved by combining several control methods after identifying the nematodes in the field and understanding their biology.

Chapter 2 - Early history and significance of the peanut crop is discussed. Annual world production of peanuts at 30 million tons makes this crop one of the most important agricultural commodities. Unusual physiology, inflorescence, and infructescence of the peanut plant make it an attractive object for scientific research. Many physiological processes in the peanut plant, such as vegetative growth, nitrogen fixation, and fertilization are associated with light. Present research revealed high light-sensitivity of the spermidines from the peanut reproductive organ, keel. A new spermidine triamide derivative has been isolated from peanut flowers and identified as N^1-acetyl-N^5, N^{10}-di-*p*-(*EE*)-coumaroylspermidine on the basis of detailed analysis of NMR, MS, and UV data. Two other spermidine conjugates, N^1, N^5, N^{10}-tri-*p*-(*EEE*)-coumaroylspermidine and di-*p*-(*EE*)-coumaroylspermidine, as well as four flavonoid conjugates (quercetin-3-glucoside, quercetin-3-glucuronide, isorhamnetin-3-glucoside, and isorhamnetin-3-glucuronide) that have been previously reported in organs of other plants, have been found in this study in peanut (*Arachis hypogaea* L.), a representative of the Leguminosae family, for the first time. The dynamics of photoisomerization in the spermidine conjugates have been investigated. These flavonoid and spermidine conjugates may play protective roles against pests. New spermidine conjugate may be involved in regulating the plant reproduction process.

Chapter 3 - The elucidation of the contamination way of the toxigenic fungi and mycotoxins in peanuts focusing the environmental and field conditions is very important, so as to develop efficient control strategies. The peanuts "in natura" and your derived products are used in human and animal feed, besides the high frequency of toxigenic fungi like *Aspergillus flavus* and *Fusarium verticillioides* in Brazil and the importance of studies regarding mycotoxins motivated this work. Under these considerations, the present study aimed to: 1. Identify the mycoflora of the soil, air, flower, pegs, peanut kernels and hulls in the different maturation stages and after drying; 2. Determinate the occurrence of aflatoxins B_1, B_2, G_1 and G_2, cyclopiazonic acid (CPA) and fumonisins B_1 and B_2 in peanut kernels and hulls in different maturation stages and after drying.

The mycoflora study of the peanut kernels and hulls showed that *Fusarium* spp. and *Aspergillus flavus* were the most frequent fungi isolated. The presence of *A. flavus* also was detected in soil and air samples. *A. parasiticus* was isolated from the peanut kernels only after drying. The toxigenic potential tests of isolated strains demonstrated that 85.7 % of *A. flavus* analyzed showed the ability to produce aflatoxins, 38.0 % also produced CPA, 33.3 % produced both toxins and 4.7 % produced only CPA. It was observed that all *A. parasiticus* strains produced aflatoxins.

Aflatoxins and CPA were detected in 32 % of the peanut kernel samples in concentration that ranged from 4.20 μg/kg to 198.84 μg/kg and from 260 μg/kg to 600 μg/kg, respectively. These mycotoxins were detected from pod filling stage. Only aflatoxins were detected in the peanut hulls (24% of the samples) and the concentration ranged from 5.76 μg/kg to 218.52 μg/kg. In despite of two toxigenic *F. verticillioides* strains were isolated from peanut kernels and hulls, there were no fumonisins in these samples. Good agricultural practices are indicated to minimize the risks in this region and to avoid the hazards of mycotoxin contamination of peanut kernels, because the presence of toxigenic fungi *A. flavus* and *A. parasiticus* in soil.

Chapter 4 - The dairy industry encounters important economical losses and health risks due to harmful spore-forming bacteria. Contamination occurs mainly at the milking stage and microbial numbers grow during conservation before pasteurisation or sterilisation. Some of the contaminants produce degrading enzymes such as lipases and proteases and render the milk unsuitable for human consumption. Others produce toxins and cause severe food illnesses and/or food poisoning. Spore-forming bacteria can survive heat treatments and jeopardize further production processes. Sources of the bacteria are diverse: feed and faeces are paramount, but air, grass, straw, milking equipment and the farmer's skin have been mentioned as well. Logically, the bacterial diversity of a contamination source will influence the resulting bacterial diversity of raw milk. Evidence suggests that differences in season, housing strategy and feed formulation contribute to the bacterial population of milk. Seeing that the operational management on dairy farms determines many of these factors, it seems obvious that this will also determine the bacterial diversity in the produced milk. To test this hypothesis, the diversity of aerobic endospore-forming bacteria in milk obtained from five conventional and five organic dairy farms in Flanders (Belgium) was compared. Although overall diversity was highly similar, some differences, possibly correlated to operational management, were observed. The greater number of thermotolerant bacteria in milk from conventional dairy farms for example, could be correlated with dietary differences. Cows from organic farms are fed with locally cultivated crops, while cows from conventional dairy farms consume commercially available feed concentrate, which is partially composed of pelleted (and hence heat-treated) and/or imported tropical ingredients. The observed higher occurrence of *Bacillus cereus* - a soil contaminant - in milk from organic dairy farms could be due to differences in housing strategy. Current evidence indicates that differences in operational management might influence the microbial diversity of raw milk, and as a consequence might have an impact on related economical losses and health risks. Further research in this field is needed and should include other bacterial contaminants and other countries.

Chapter 5 - Bacterial functionality, as this was described by the utilization of substrates in the GN Biolog plates, was studied in different combinations of disinfection (chemical, biological, without disinfection) and cultivation (organic vs conventional) since both are common practices in agroecosystems. Within an area of 3000 m^2, which was conventionally cultivated with *Phaseolus vulgaris* (L.) for several years, a field plot experiment was set consisting of 18 plots (10x10 m each). Six plots were disinfected chemically with metham sodium, six were disinfected with organic additives (neem_cake and a mixture of basil oil, marigold and clove oil) while in the rest six plots no disinfectants were applied. Three plots from each disinfection type were cultivated organically receiving cow manure and the other three were cultivated conventionally. All plots were sowed by *Phaseolus vulgaris*. Bacterial functionality was examined in six treatments: chemical disinfection-organic cultivation (CD-O), biological disinfection-organic cultivation (BD-O), without disinfection–organic cultivation (WD-O), chemical disinfection-conventional cultivation (CD-C), biological disinfection-conventional cultivation (BD-C) and without disinfection-conventional cultivation (WD-C). Soil samples were collected at two sampling occasions: 1) at the flowering stage and 2) one month after the harvest of plants' aboveground biomass.

Shannon diversity index showed no response either to treatment or to sampling date reflecting the elimination of quantitative differences in bacterial functional diversity. On the contrary when the parameters K (maximum absorbance) and r (rate of changes from zero to K), which produced by the logistic curves fitted to the optical density data, were analysed, samplings and treatments were clearly separated. Since the distinction between treatments referred only to the degree of the utilization of some of the substrates, the differences in bacterial functionality were qualitative and not quantitative. The substrates responsible for the discrimination between treatments corresponded only to the 8-12% of the total_number of substrates used. It seemed that either cultivation or disinfection were disturbances of low severity for the bacterial functionality. The type of disinfection seemed to have a greater influence in conventional cultivated plots, by creating more distinct groups of samples, than in organically cultivated ones. We could hypothesized that the presence of this legumes species for several years in the study area, resulted in the dominance of a specific soil bacterial community, so the effects of the applied disturbances were of less importance for its function compare to those induced by the presence of this plant species.

Chapter 6 - Peanut (*Arachis hypogaea* L.) is a regional crop of Argentina. It is grown in the centre-south of Córdoba province, where almost 96% of the national production and all of its industrial processing are concentrated (Fernandez and Giayetto, 2006). Peanut is an important subsistence and cash crop in the semi-arid tropics where it often suffers from drought or saline stress. Little is known about the molecular events involved in its adaptive responses to stress and about the biochemical and physiological basis of tolerance by rhizobia nodulating peanut roots and their symbiosis. The aim of the present chapter is to contribute to the knowledge of physiological and biochemical response (trehalose content; lipids and periplasmic glucans) of peanut microsimbionts under different types of stress. This response will be compared between rhizobia, and it probably will contribute in the selection of a great tolerant strain for future essays of inoculation.

Results indicate that there are different responses of peanut-rhizobia to abiotic stress, indicating genetic variability in tolerance to these environmental constraints. Trehalose accumulation is strain dependent. However other parameters were similar in different strains. Thus the polymerization degree of periplasmic glucans was affected under the stress conditions applied; this was evidenced in its molecular weight profile. Also phospholipids increased under hypersaline and hyperthermic conditions.

Results suggest that periplasmic glucans and phospholipids would be implicated in the adaptive response of peanut strains. Further analysis of these molecules in peanuts rhizobia must be carried out to select bacteria tolerant to stress for inoculants formulations.

Chapter 7 - This paper discusses the concept of scientific paradigms and draws parallels with paradigms in agriculture, and in particular focuses upon organic farming and genetic engineering. There is anecdotal evidence that they are both potential new agricultural paradigms. Different writers have referred to both as new ‘paradigms’ in agriculture, though the qualitative evidence is stronger for organic farming than genetic engineering. The attitudes of agricultural professionals are important as they are responsible for creating, developing and diffusing innovations as well as informing and educating farmers and the public. This paper summarizes past research on the attitudes of Australian agricultural professionals (n=185) towards organic farming (Wheeler 2008) and genetic engineering (Wheeler pending). These two sets of results are compared to assess if professional attitudes to the two innovations are driven by similar influences or are quantitatively significantly different. In addition, analysis using bivariate ordered probit regression is conducted in this study on the nature of professionals’ answers to the net benefits of the two innovations themselves. Do professionals tend to be negative or positive about both innovations, or positive about one and negative about the other? From a comparison of the results, it is shown there is some quantitative evidence that organic farming represents a potential new sub-paradigm in agriculture, but only very weak evidence that genetic engineering is a potential new paradigm.

Chapter 8 - The number and popularity of farmers' markets in the United States have grown concurrently with organic production and consumer interest in locally and organically produced foods. This research describes the significance of farmers' markets as market outlets for many organic farmers, and recent shifts in relationships between organic growers, market managers, and customers. Market managers in more than 20 States answered questions by phone pertaining to the 2002 market season. Their responses provide insight into recent grower, manager, and customer decision making and attitudes about foods advertised and sold as organic at farmers' markets. Specific areas of interest include participation of organic farmers in markets, demand for organic products, advertising strategies used to highlight organic products, and price premiums charged for organic products.

In: Organic Farming and Peanut Crops
Editors: D. C. Grossman, T. L. Barrios, pp. 1-64
ISBN: 978-1-60876-187-6

Chapter 1

NEMATODE CONTROL IN ORGANIC FARMING SYSTEMS: FROM THE LABORATORY TO THE FIELD

Yuji Oka
Nematology Unit, Gilat Research Center, M. P. Negev 85280, Israel.

ABSTRACT

Control of plant-parasitic nematodes is difficult, especially in organic farming systems, because in contrast to foliar diseases and insect pests, effective control methods and monitoring systems are not always available. Organic farmers struggle with nematode problems using cultural, physical and biological control methods. Cultural practices such as crop rotations, use of resistant cultivars, and soil amendments, are very commonly integrated into the soilborne-disease control strategies in organic farming systems. Nematode-resistant varieties are available mainly for root-knot (*Meloidogyne* spp.) and cyst (*Heterodera* and *Globodera* spp.) nematodes in a few crops. Use of grafted vegetable seedlings has been become popular in Solanaceae and Cucurbitaceae crops, but no rootstocks resistant to root-knot nematodes are commercially available for the latter. Soil amendments, especially those with high nitrogen-to-carbon ratios, have been reported to exhibit nematicidal and fungicidal activity, mainly through the release of ammonia during their decomposition in the soil and/or through the consequent increased populations of antagonistic microorganisms. Nematode control efficacy of ammonia can be enhanced by manipulating soil environments, such as soil pH and temperature, and by suppressing microorganisms involved in nitrification. Plants that possess nematicidal compounds or their precursors are often incorporated to control soilborne diseases (biofumigation), with a special focus being placed on members of the Brassicaceae family. This effect is generally attributed to biocidal isothiocyanates, which are released from glucosinolates in plants by enzymatic activity. Other plants containing nematicidal compounds are useful as cover crops and/or green manure. The effects of composts on plant-parasitic nematodes have not been studied as thoroughly as their effects on fungal diseases. Soil solarization has been developed as a physical control method, and used to control soilborne diseases and weeds. This control method is also effective in nematode control; however, combinations of soil solarization and organic amendments greatly improve the nematode-control efficacy of soil solarization alone. Use of biological agents in nematode control is not as popular or successful as it is in insect control although many nematode-antagonistic organisms are known. Effective management of plant-parasitic nematodes can be achieved by combining several control methods after identifying the nematodes in the field and understanding their biology.

INTRODUCTION

Worldwide concern for the environment and food safety has resulted in restrictions on the use of chemical pesticides. Nematicides, including soil fumigants, may be the most risky pesticides in agriculture because they are applied to the soil at high rates, and have the potential to leach to the groundwater, or be emitted to the atmosphere. In fact, several nematicides have been removed from the market for these very reasons. Sustainable agriculture, especially organic farming, has been redeveloped to solve such problems as well. However, a continually increasing world population requires an increasing food supply, and agrochemicals, including fertilizers, continue to play an important role in crop production. This may be the "conflict" of modern agriculture today, but solutions will be found by continual cooperation among scientists, researchers, extension services, farmers and consumers. Nematodes are considered to be relatively "new" and minor plant pathogens as compared to insects and fungal diseases, but the crop damage caused by nematodes may not be correctly estimated due to their main habitat (soil) and their infection sites (roots) in plants. The number of available nematicides has decreased in conventional agriculture due to their removal from the market and almost no release of new products: this is probably due to the technical difficulties and economic considerations involved in new nematicide development for agrochemical companies. In this situation, alternative non-chemical control methods have also become important in conventional agriculture. Satisfactory control of plant-parasitic nematodes in organic farming systems requires a broad knowledge of nematode biology and control methods because unlike soil fumigants in conventional agriculture, there is no single method that controls all nematode species on all crops. In this chapter, methods for controlling nematodes without the use of synthetic chemicals are described. Some of these methods may not be acceptable or meet the rules of organic farming authorities in some areas; others may have only been tested in small experimental units or the laboratory. Nevertheless, they are included here to provide some ideas and theories for the development of new control methods. In most cases, the details of the various control methods, such as application rates, protocols, cultivar names and nematode species, are not provided. Cited references may be useful in this regard, and some provide practical applications in the field.

Plant-parasitic nematodes

General description of nematodes: Nematodes (also called eelworms or roundworms), which belong to the phylum Nematoda, are the most abundant multicellular animals on our planet. The phylum consists of about 20,000 highly diverse species. Free-living nematodes that feed on bacteria and fungi can be found in the soil, fresh water and marine water. Predatory nematodes feed primarily on prey nematodes. Parasitic nematodes are found in plants, arthropods, mollusks and animals, including humans (e.g. *Ascaris lumbricoides*). Plant-parasitic nematodes are generally microscopic (ca. 0.3-4.0 mm), whereas many species of animal-parasitic nematodes are visible to the naked eye. For instance, a parasitic nematode of whales can reach as long as 13 m. A large proportion of all nematode species are free-living (feeding on microbes), and those that exist in the soil have important roles in the agroecosystem, such as nutrient turnover. The most famous free-living nematode is *Caenorhabditis elegans*, which is used as a model organism in molecular and developmental biology. The nematode was the first multicellular organism to have its entire genome sequenced. Nematodes are non-segmented and bilaterally symmetric with a complete digestive system (Fig. 1). They have no circulatory or respiratory systems. Nematodes reproduce in a variety of ways, including sexually (amphimixis), via self-fertilization (hermaphroditic) and asexually (parthenogenesis). Plant-parasitic nematodes are translucent, and have a vermiform shape when they hatch from eggs (Figs. 2A, 3A). However, their body shape may change when they develop into adults.

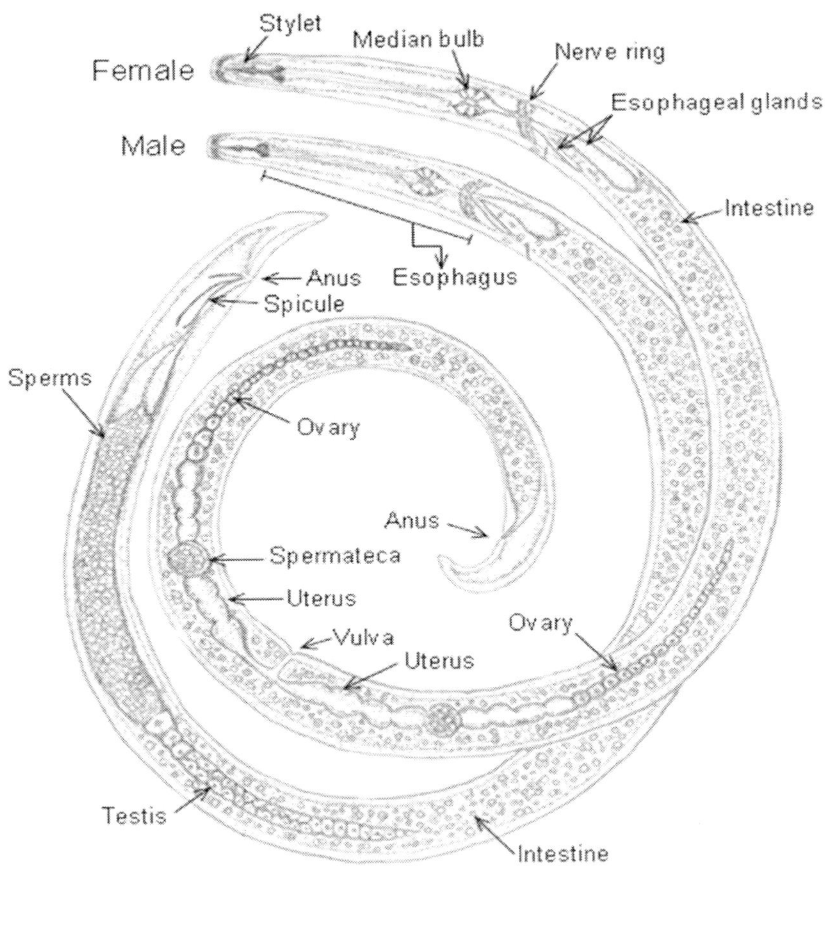

Fig. 1. Body structure of typical female and male plant-parasitic nematode.

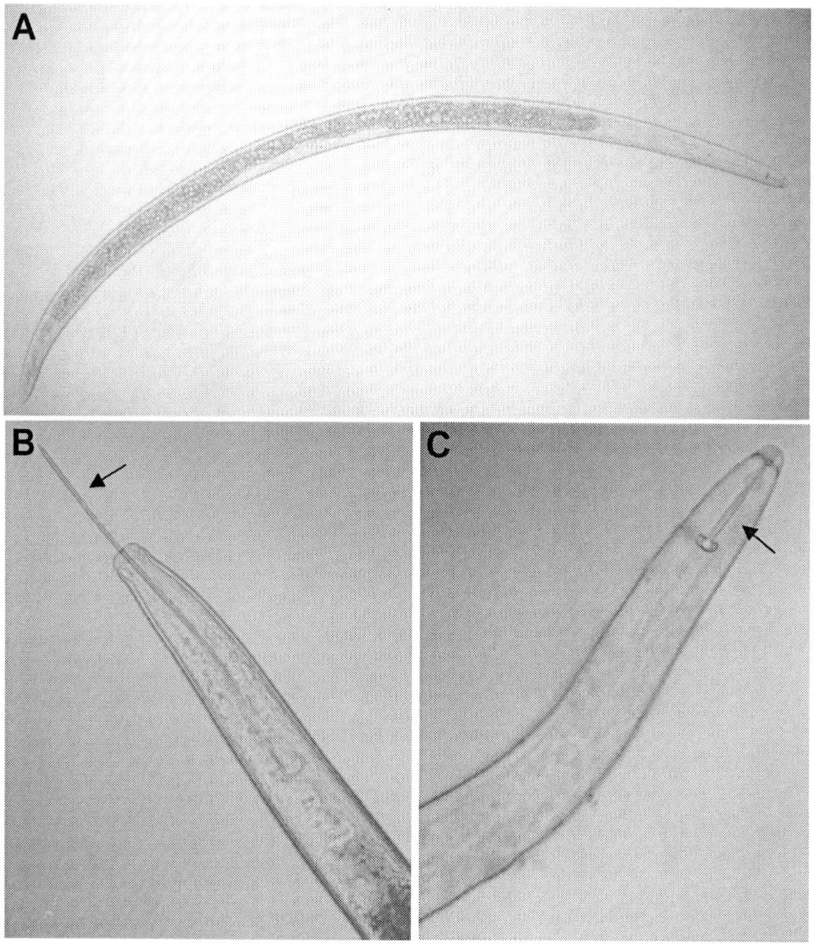

Fig. 2. Plant-parasitic nematodes. A: A vermiform plant-parasitic nematode. B: Anterior part of *Longidorus* sp. (order Dorylaimida) C: Anterior part of *Helicotylenchus* sp. (order Tylenchida). Stylets are marked with arrows.

Plant-parasitic nematodes: More than 4,000 nematode species are plant parasites. All plants, including vegetables, field crops, flowers, weeds and trees, can be attacked by at least one plant-parasitic nematode species. Fungi, including cultivated edible mushrooms, are also targets of fungivorous (fungal-feeding) nematodes. Some plant-parasitic nematodes can also complete their life cycle by feeding only on fungi. Most species of plant-parasitic nematodes attack only the root system; however, some species can attack aboveground parts, including stem, leaf, buds and seeds (Manzanilla-Lopez *et al.*, 2004). Plant-parasitic nematodes can be divided into two main groups according to their mode of parasitism: endoparasites and ectoparasites (Table 1). Endoparasites penetrate the plant tissue (Fig. 3B,C): their body is partially (semi-endoparasites) or entirely embedded in the plant, they take up nutrients via a stylet from live cells and they develop to adulthood inside the plant or on the plant surface.

Table 1. Major nematode species and their symptoms at infection sites and in major host plants

Sedentary endoparasites and semi-endoparasites*

Causal nematode	Common name	Symptom (at infection site)	Main host plant (showing the symptom)
Globodera rostochiensis, G. pallida	Potato cyst nematodes	Cysts on roots	Potato, tomato, eggplant
Heterodera avenae	Cereal cyst nematode	Proliferation of lateral roots, cysts on swollen roots	Wheat, barley, oat
Heterodera glycines	Soybean cyst nematode	Cysts on roots	Soybean, various beans
Heterodera schachtii	Sugar beet cyst nematode	Cysts on lateral roots, proliferation of lateral roots	Sugar beet, spinach, cabbage, rapeseed, celery, carrot, radish
Meloidogyne spp.	Root-knot nematodes	Root galls	Many
Meloidogyne spp.	Root-knot nematodes	Cracks	Sweet potato
Meloidogyne spp.	Root-knot nematodes	Warts, scabs, swelling	Potato
Meloidogyne hapla	Root-knot nematode	Excessive root branching, small galls	Many
Naccobus spp.	False root-knot nematodes	Bead-like root galls	Potato, Solanaceae, crucifers and legumes
Rotylenchulus spp.	Reniform nematodes*	Small sand-like bodies (females and egg masses) on root surface	Cotton, soybean, pineapple, various legumes
Tylenchulus semipenetrans	Citrus nematodes*	Encrusted appearance, (females and egg masses) on root surface, discoloration	Citrus, grapevine, olive, persimmon

Migratory endoparasites

Aphelenchoides spp.	Foliar nematodes	Angular leaf spots, interveinal necrosis, misshapen, white tip	Several (ornamentals, strawberry, mushroom, rice, peanut)
Brusaphelenchus spp.	Pine nematodes	Wilt and death	Pine trees
Ditylenchus spp.	Bulb and Stem nematodes	Leaf distortion, leaf and stem swelling, lesions, short internodes, twisting	Foliage of several plants (ornamentals, strawberry, legumes, cereals)
Ditylenchus spp.	Bulb and Stem nematodes	Internal ring-rot, discoloration, swollen scales, dry rot	Bulbs (ornamental, garlic, onion), tubers (potato)

Table 1. Major nematode species and their symptoms at infection sites and in major host plants (continued)

Migratory endoparasites

Helicotylenchus spp.	Spiral nematodes	Longitudinal necrotic lesions	Many plants (banana, sugarcane)
Hoplolaimus spp.	Lance nematodes	Necrotic lesions	Several (cotton, soybean, grasses)
Pratylenchus spp.	Lesion nematodes	Longitudinal necrotic lesions, color lesion, root-rot	Many
Radopholus spp.	Burrowing nematodes	Longitudinal necrotic lesions, root-rot, toppling	Many (banana, citrus, ornamentals)

Migratory and sedentary* ectoparasites

Belonolaimus spp.	Sting nematodes	Stubby roots, shrunken lesions,	Many (cereals, maize, grasses, cabbage)
Criconemella spp.	Ring nematodes*	Lack of feeder roots, lesions	Several (stone fruits)
Helicotylenchus spp.	Spiral nematodes	Root lesion	Many (banana, sugarcane)
Longidorus and *Paralongidorus* spp.	Needle nematodes	Root tip galls	Many
Paratylenchus spp.	Pin nematodes	Necrotic lesions	Many
Trichodorus and *Paratrichodorus*	Stubby root nematodes	Stubby roots	Many
Tylenchorhynchus	Stunt nematodes	Short, discolored and sparse roots	Many
Xiphinema spp.	Dagger nematodes	Dagger roots, root tip galls	Many (grapevine)

Ectoparasites stay in the soil (rhizosphere) and take up nutrients by piercing root cells with their stylet. Nematodes can also be divided into sedentary and migratory parasites (Table 1). Sedentary nematodes stop moving after they start feeding inside plants (sedentary endoparasites), whereas migratory nematodes move inside the plant tissue (migratory endoparasites) or in the rhizosphere (migratory ectoparasites), feeding from cell to cell. Sedentary ectoparasites, such as the ring (*Criconemella* spp.) nematode, may feed from a single site or plant cell for a prolonged period of time while remaining outside the roots. Important sedentary endoparasites are root-knot (*Meloidogyne* spp.), cyst (*Globodera* and *Heterodera* spp.) and reniform (*Rotylenchus* spp.) nematodes; important migratory endoparasites are lesion (*Pratylenchus* spp.), burrowing (*Radopholus* spp.), spiral (*Helicotylenchus* spp.) and lance (*Hoplolaimus* spp.) nematodes, while important migratory ectoparasites include stubby-root (*Paratrichodorus* and *Trichodorus* spp.), dagger

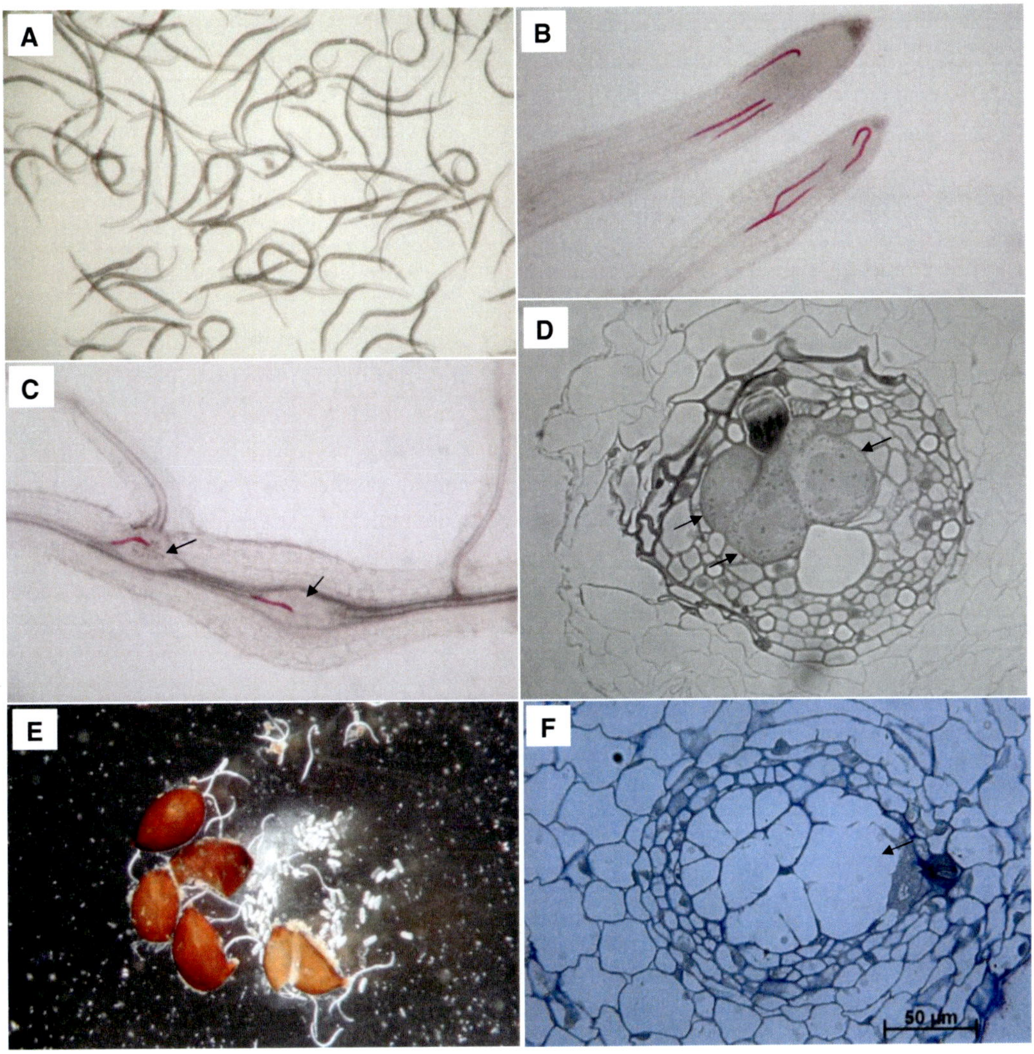

Fig. 3. Infection process of *Meloidogyne* spp. (A-D) and *Heterodera avenae* (E-F). A: Infective second-stage juveniles of *M. javanica*. B: *M. javanica* juveniles in root tips. C: Juveniles feeding on giant cells (arrows) in tomato root. D: Microsection of the feeding site of *M. marylandi* in wheat root. Note the multinuclear giant cells (arrows) in the vascular cylinder. E: Cysts, hatched juveniles and eggs. F: Syncytium (arrow) in wheat root.

(*Xiphinema* spp.), needle (*Longidorus* and *Paralongidorus* spp.), stunt (*Tylenchorhynchus* spp.) and pin (*Paratylenchus* spp.) nematodes. *Helicotylenchus* spp. can also be migratory ectoparasites, depending on the species and the plant host (Whitehead, 1997; Hussey & Grundler, 1998; Shurtleff & Averre, 2000; Manzanilla-Lopez *et al*., 2004).

All plant-parasitic nematodes use their stylets for feeding (Fig. 2B,C). They use it to pierce plant cells, inject a secretion to ease subsequent ingestion of cytoplasm, and attack another cell after feeding. However, some nematodes have a more complicated mode of feeding. For example, root-knot nematodes form very specific feeding cells, which are called "giant cells", in the host plant's roots (Fig. 3C,D). The nematode secretion injected into certain cells (generally pericycles in the vascular cylinder) causes proliferation of nuclei without mitosis, enlarges the cell volume (hypertrophy), enriches organelles and increases the cell's metabolism (Jones & Northcote, 1972). Root-knot nematodes feed on a few giant cells throughout their lives without moving in the root (sedentary endoparasite), developing into adults there. The nematode secretions are thought to be essential for the function of giant cells. Cyst nematodes (*Heterodera* and *Globodera* spp.) induce another type of feeding cell called "syncytium" which is similar to the giant cell (Fig. 3E,F). The syncytium is also a large multi-nuclear cell, but it is formed by the fusion of several cells around the initial syncytial cell—the first cell to receive the nematode's secretion, by cell-wall degradation (Golinowski *et al.*, 1997).

Estimated crop losses due to plant-parasitic nematodes amount to over $77 billion worldwide (Sasser & Freckman, 1987). The most crop-destructive nematode genera may be the root-knot nematodes (*Meloidogyne* spp.), cyst nematodes (*Heterodera* and *Globodera* spp.) and lesion nematodes (*Pratylenchus* spp.), especially on vegetable crops; however, nematode species associations with crops vary, depending on the region and the crop. Aside from the direct damage to plants caused by nematode infection, some nematodes transmit plant viruses (Brown *et al.*, 2004), or the nematode infection can synergistically enhance the severity of plant diseases caused by other plant pathogens. The most famous example of such interactions is that between the root-lesion nematode *Pratylenchus penetrans* and verticillium wilt (*Verticillium dahliae*) on potato (Martin *et al.*, 1982).

Symptoms: Plants infected with nematodes on their roots show symptoms similar to those caused by drought and nutrient deficiencies, such as chlorosis, stunting and wilting due to disrupted nutrient and water uptake and utilization in plants (Figs. 4A,B, 5A). If such symptoms cannot be corrected by fertilization and/or irrigation, nematode infection is suspected. In the field, such symptoms are generally observed as circular or elongated patches along the direction of cultivation. However, symptom development is affected by nematode density, the crop's susceptibility to nematodes and environmental conditions. Under high nematode densities, plant growth may be inhibited in its early stages, i.e. shortly after transplanting or seeding. Under light infestation levels, aboveground symptoms may appear later. Most plant-parasitic nematodes infect plant roots, but the morphologies (symptoms) on roots infected with nematodes vary with nematode species, such as necrotic lesions and root rot, root galling, stunted or stubby roots and abnormal proliferation of lateral roots (Shurtleff & Averre, 2000; Manzanilla-Lopez *et al.*, 2004) (Table 1) (Figs.4C-F, 5B). Symptoms on aboveground parts which have been directly infected with nematodes include leaf lesions and rot, browning of the internal stem, discoloration, distortion, proliferation of lateral buds, and seed and leaf galls (Table 1).

Diagnosis: The first described plant-parasitic nematode was *Anguina tritici*, which causes galls on wheat seeds and was described in 1743 by Turbeville Needham. Plant-parasitic nematodes can be isolated from soils and/or plant tissues by means of simple equipment, such as funnels and sieves (Hooper & Evans, 1993; Shurtleff & Averre, 2000). However, identification of plant-parasitic nematodes among the isolated nematodes is difficult for untrained personnel because of their microscopic size and the abundance of non-parasitic

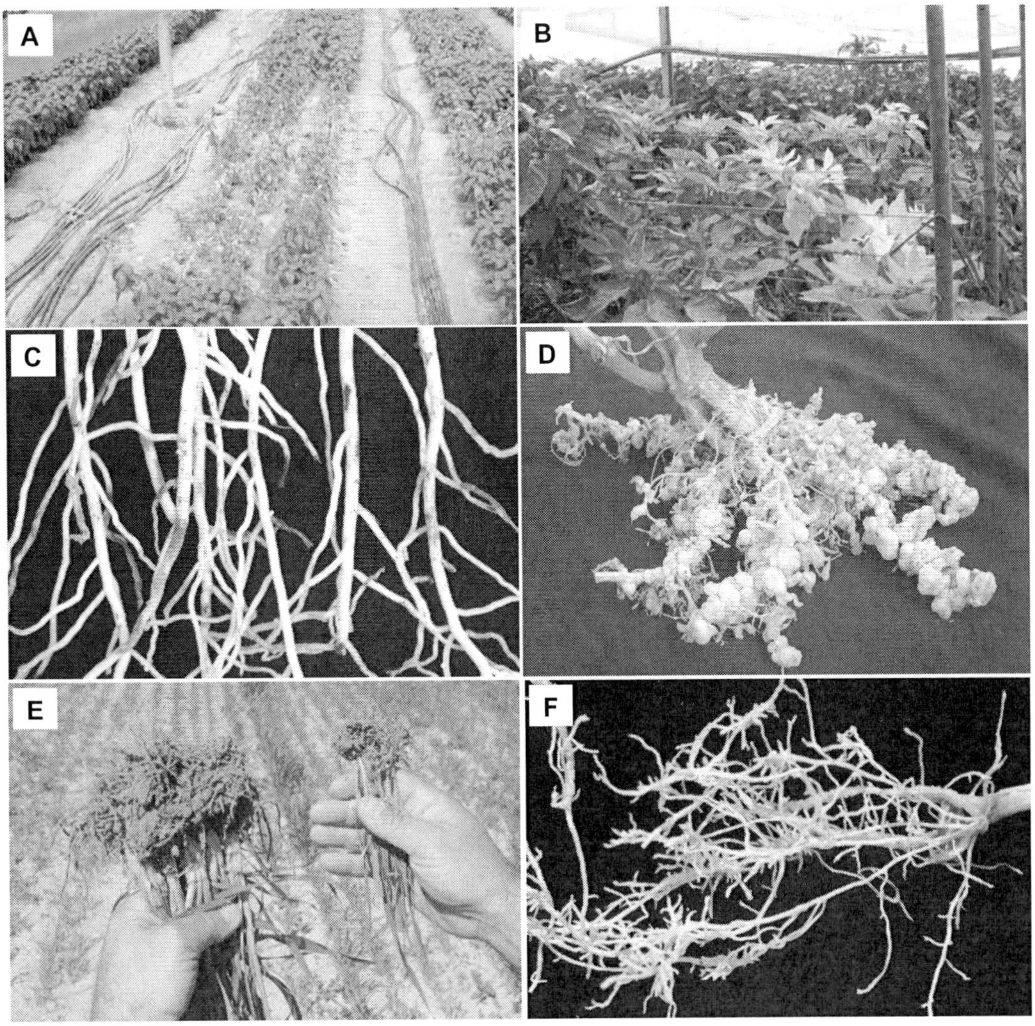

Fig. 4. Symptoms of nematode infection. A: Stunting and wilting basil plants infected with the root-knot nematode *Meloidogyne javanica*. B: Chlorosis on pepper plants infected with the root-knot nematode *M. incognita*. C: Lesions caused by lesion nematode *Pratylenchus* sp. on amaryllis roots. D: Galled roots of cucumber caused by *M. javanica*. E: Healthy (left) and *Heterodera avenae*-infected (right) wheat root. F: Lateral root proliferation and swellings on *H. avenae*-infected wheat.

(free-living) nematodes in the sample. Identifying symptoms on plants infected by certain nematodes is much easier than identifying nematodes from samples. Root galls caused by root-knot nematodes (*Meloidogyne* spp.) are easily diagnosed, although other nematodes, such as dagger nematodes (*Xiphinema* spp.) and false root-knot nematodes (*Naccobus* spp.), certain bacteria (*Agrobacterium*) and the fungus *Plasmodiophora brassicae* cause similar root galls on a number of plant species. Adult females of cyst nematodes (*Heterodera* and *Globodera* spp.) can be seen with the naked eye on root surfaces. The best way to diagnose suspected plants or soils is to send the samples to authorized laboratories. Sampling methods are also important for accurate diagnosis (Shurtleff & Averre, 2000). Recent progress in diagnosis by molecular methods has enabled the identification of even single nematodes.

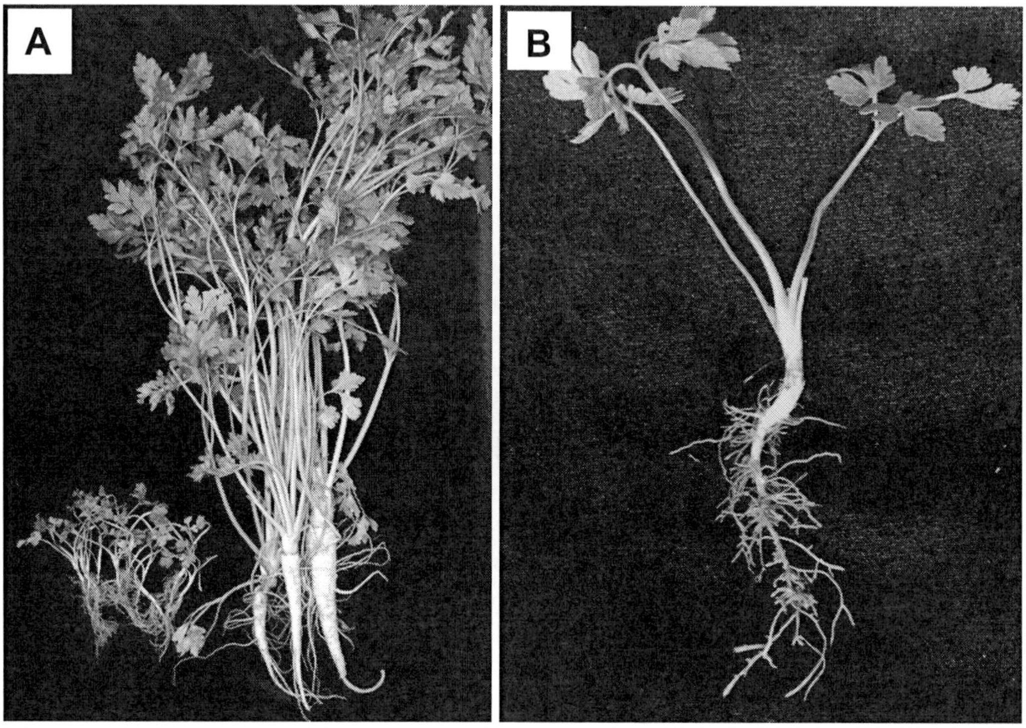

Fig. 5. Symptoms of nematode infection on parsley. A: Healthy (right) and *Longidorus* sp.-infected (left) plants. B: Plant infected with *Longidorus* sp.

Diagnostic kits for easy and rapid nematode identification are under development and are expected to be available for use in the near future.

Nematode control in conventional agriculture: Nematodes have been controlled by soil fumigants and/or non-fumigant nematicides, mainly in high-value crops, for more than 50 years. Nematicide application is relatively straightforward and does not require nematode identification or an understanding of its biology. The first chemical used for nematode control was carbon disulfide, mainly to control phylloxera in vineyards. This chemical was used to control the sugar beet cyst nematode *Heterodera schachtii* in 1811, and later, root-knot nematodes (Bessey, 1911). Later still, several soil fumigants were developed, including chloropicrin (used as tear gas in World War I), ethylene dibromide (EDB), dibromochloropropane (DBCP), methyl bromide, the dichloropropane-dichloropropene mixture (D-D), and several methyl isothiocyanate (MIT)-generating compounds, such as metham sodium. EDB, D-D and DBCP were used mainly for nematode control, whereas others were general soil fumigants for the control of weeds, fungi and insects which also had nematicidal activity. Most of these fumigants were developed from the late 19th century to mid 20th century (Rich *et al.*, 2004). Following their application to the soil, the fumigants' gas phase spreads between the soil particles, contacts the nematodes and kills them. The increased yields resulting from soil fumigation led researchers to understand the importance of nematodes' impact on crops, and more research on these organisms began. A large proportion of these fumigants (EDB, DBCP and D-D) were subsequently banned due to their toxicity to the environment and to humans. Methyl bromide, which served as the main soil fumigant for decades, was phased out in 2005 in most developed countries, and its phaseout in developing countries is planned for 2015. Several formulations of 1,3-dichloropropene (1,3-D), which is the nematicidal component in D-D, with chloropicrin and/or MIT-generators have replaced methyl bromide; however, their efficacy is generally lower, and their use (especially for 1,3-D) has been restricted in some areas. These soil fumigants are used before seeding or transplanting because they are phytotoxic, except for DBCP, which was also used after planting in several crops, but is no longer allowed.

Non-fumigant nematicides which, like some insecticides, are carbamates or organophosphates were developed after 1960, including thionazin, fensulfothion, aldicarb, fenamiphos, ethoprophos, oxamyl, terbufos, carbofuran, cadusafos and fosthiazate (Rich *et al.*, 2004). These nematicides can be used before and after planting because they are not generally phytotoxic. They are applied to the soils as granules or in solution, and the active compounds spread through the soil with the soil water. These nematicides are neurotransmission-inhibitors: they do not kill nematodes at the recommended application doses in the field, they paralyze them. When the nematicides are removed by water or degradation in the soil, the paralyzed nematodes may recover and become motile, and may also be infective. These non-fumigant nematicides are also highly toxic to humans. The median lethal dose (LD_{50}) of such nematicides in rats via oral administration is between 0.5 and 73 mg a.i./kg body weight. Because of their high toxicity, their use has been restricted and some have been removed from the market. Currently, only a very small number of nematicides are commercially available, and in contrast to fungicides and insecticides, new nematicides have barely been developed in the last decade. Therefore, alternative nematode control methods to chemical nematicides and soil fumigants are also being sought for conventional agriculture, and the nematode control methods used in organic farming systems have been integrated into nematode control strategies for conventional agriculture.

Nematodes in organic farming

Generally speaking, soilborne diseases, including those caused by nematodes, are thought to be less frequent in well-managed organic farming systems than in conventional agriculture, due to the former's emphasis on creating fertile and healthy soils by applying organic amendments, such as green manures and compost (Griffith *et al.*, 1994; Ferris *et al.*, 1996; Berkelmans *et al.*, 2003; van Bruggen & Termorshuizen, 2003). In such soils, a wide variety of microorganisms, including natural enemies and antagonists of nematodes, play an important role in maintaining the microbial balance in the soil ecosystem (Bengtsson *et al.*, 2005; Hole *et al.*, 2005). Nematode species that are present in the soil may serve as an indicator of soil health (Neher, 2001). However, the statement that "organic soils don't have nematode problems" is not always true; higher population densities of lesion nematodes (*Pratylenchus* spp.) have been found in organic soils than in a conventional system (Neher, 1999). In Germany, plant-parasitic nematodes have become a serious problem in organic farms—root-knot nematodes (*Meloidogyne* spp.) and lesion nematodes (*Pratylenchus* spp.) in vegetable crops, and cereal cyst nematodes (*Heterodera* spp.) and *Pratylenchus* in cereals (Hallmann *et al.*, 2007). This may be due to the absence of nematode-control practices, including black fallow and crop rotations with non-host-, antagonistic- or trap-plants of nematodes.

If nematode infestation is suspected based on the above-described symptoms, plant and/or soil samples should be sent to a plant disease diagnostic laboratory. Consulting an extension service is also recommended. Samples should be collected according to the methods dictated by the laboratory. Generally, soils from the root zone and roots are collected for nematode isolation (Shurtleff & Averre, 2000). For comparison, soil samples from healthy plants in the same field may be useful. Identifying nematode species, not only genera, may be more important in organic farming systems than in conventional farming, because control methods may vary depending on the nematode species in the former. In contrast, nematicides or soil fumigants in conventional farming control nematodes almost independently of nematode species. Once a nematode species or genus is known, control measures can be selected according to its biology, host plants, availability of resistant varieties, etc.

Nematode control methods in organic farming

From ancient times, humans have struggled with plant diseases without the availability of synthetic pesticides, and even with no awareness of the causal plant-pathogens and control mechanisms in their cultural practices (Thurston, 2004). Crop rotation, including fallow, is one example of a method which avoids the build-up of certain pathogens in the soil by changing crops from season to season. This method can probably also avoid the depletion of certain nutrient elements by one crop if it is grown for several seasons in the same field. However, returning to cultural practices used in the era before development of agrochemicals cannot supply organic products that meet today's demand for quality and quantity. More logical, sophisticated and science-based control strategies, in addition to traditional control methods, should be taken to ensure satisfactory yields. Here, nematode control by cultural, physical, "chemical", and biological methods and integrated management of nematodes are mentioned. The term "management" is generally used to refer to the application of several methods for the whole agrosystem to minimize economic losses (ideally below economic threshold) caused by a certain pest, in our case nematodes. In contrast, "control" is a specific act designed to reduce number of pathogens, in our case nematodes. In organic farming, "nematode management" is conducted by combining several control methods.

Cultural control

1. Sanitation

Sanitation is the most effective means in nematode management of preventing their invasion of the farm. Once a nematode has invaded and established itself in a field or greenhouse soil, its eradication is very difficult, if not impossible. Propagation materials, such as seeds, tubers, bulbs and seedlings, should be free of nematodes and diseases. Such materials must be purchased from trustworthy suppliers or nurseries. Transplanting seedlings, especially of trees, which have been grown in nematode-infested soil represent a very common route for nematode distribution to fields. If there is any suspicion of infection, a sample should be sent to an authorized laboratory before planting. Some methods can be applied to control nematodes in infected planting materials, such as treatment with hot water, but these are not recommended for untrained growers because the treatment may also kill the planting materials or may not kill the nematodes effectively.

If nematodes are known to exist in a certain field, the most important practice is to prevent their spread into other fields. Their innate ability to move is very limited (less than 1 m/year), but they may spread via soil on tractor wheels or attachments (plough, discs, etc.) or on workers' boots. Washing this equipment is essential before entering another field. Nematodes also move by drainage from a nematode-infested field into other fields. It is important to prevent such drainage from entering another field, and it should not be used for irrigating other fields. Anhydrous dormant nematodes and cysts may be dispersed by winds.

In the field, nematode occurrence is generally spotty. Nematode-infected plants, including roots, should be removed from the field even before harvest to prevent their population growth in the soil, and the plant material should be burned. Many kinds of weeds and volunteer crop plants can serve as good host plants, increasing the nematode populations in the soil. Weed control is therefore a very important practice in nematode control when resistant or non-host crop varieties or crop covers are used to reduce nematode density in the soil.

2. Timing of planting or harvest

Nematode damage can be avoided or reduced by manipulating planting or harvest timing (Bridge, 1996). Nematode activities are greatly influenced by environmental conditions, such as temperature, soil moisture and root exudates. If crop plants are established before nematodes become active in the soil, crop damage by nematodes will be much lighter than when crops are planted in soil in which active nematodes already exist. For example, early seeding and supplementary irrigation of wheat reduced the damage caused by the cereal cyst nematode *Heterodera avenae*, because the dormant nematode eggs need low temperatures and high soil moisture for hatching (Mor *et al.*, 1992). Early seeding enables wheat plants to establish and develop a large root system before nematode hatch. Such plants have longer roots, and can take up more water from deep in the soil than plants seeded later. Alternatively, late planting when soil environments are deleterious for nematodes may also reduce nematode damage. Late seeding of soybean is one recommendation to avoid high densities of the soybean cyst nematode *H. glycines* at seeding, thereby reducing nematode damage (Slack *et al.*, 1972). However, this practice is not always a viable management option for nematode control (Hussey & Boerma, 1983; Todd, 1993).

3. Resistant varieties

Varieties that are resistant to nematodes, like those resistant to other plant pathogens such as fungi and viruses, are available for several crops (Starr *et al.*, 2002). Use of nematode-resistant varieties is an effective, inexpensive and easy way of protecting crops from nematode damage. Although many resistant genotypes are known among plant species, resistant cultivars are only available in a few crops and they are only resistant to only a small number of nematode species (Roberts, 1992; Peng & Moens, 2003; Starr & Roberts, 2004) (Table 2), probably because the introduction of resistance genes may be accompanied by that of undesirable crop traits. The term "resistance" to nematodes is used to describe the ability of

Table 2. Major crops for which cultivars or rootstocks resistant to plant-parasitic nematodes are available

Crop	Nematode	Nematode species
Banana and plantain	Burrowing nematode	*R. similis*
Barley	Cereal cyst nematode	*H. avenae*
Citrus rootstock	Citrus nematode	*T. semipenetrans*
Coffee rootstock	Root-knot nematodes	*M. incognita, M. exigua*
Cotton	Root-knot and reniform nematodes	*M. javanica, M. incognita, M. arenaria, R. reniformis*
Eggplant rootstock	Root-knot nematodes	*Meloidogyne* spp.
Garlic	Stem and bulb nematode	*D. dipsaci*
Grapevine rootstock	Root-knot, dagger and citrus nematodes	*M. javanica, M. incognita, X. index, T. semipenetrans*
Maize	Root-knot and spiral nematodes	*Meloidogyne* spp., *Helicotylenchus* spp.
Oat	Cereal cyst and stem and bulb nematodes	*H. avenae, D. dipsaci*
Peanut	Root-knot nematode	*M. arenaria*
Pepper and its rootstock	Root-knot nematodes	*M. javanica, M. incognita, M. arenaria*
Potato	Potato cyst, false root-knot and reniform nematodes	*G. rostochiensis, G. pallida, N. aberrans, R. reniformis*
Rice	Foliar and root-knot nematodes	*A. besseyi, M. graminicola*
Soybean	Soybean cyst, root-knot and reniform nematodes	*H. glycines, Meloidogyne* spp., *R. reniformis*
Stone fruit rootstock	Root-knot and lesion nematodes	*M. javanica, M. incognita, M. arenaria, P. vulnus*
Strawberry	Stem and bulb nematode	*D. dipsaci*
Sweet potato	Root-knot and stem and bulb nematodes	*M. incognita, D. dipsaci*
Tobacco	Root-knot and tobacco cyst nematodes	*M. incognita, M. javanica, G. tabacum*
Tomato and its rootstock	Root-knot and reniform nematodes	*M. javanica, M. incognita, M. arenaria, R. reniformis*
Wheat	Cereal cyst nematode	*H. avenae*

a plant to suppress nematode reproduction (Roberts, 2002). A resistant variety may prevent nematode proliferation but may nevertheless suffer from nematode invasion, resulting in yield reduction. An ideal resistant variety gives higher yields in nematode-infested fields than the well-accepted susceptible varieties, and gives yields similar to those of susceptible cultivars in non-infested fields. Quality is an important factor for resistant varieties as well. Nematode-resistant varieties are mainly resistant to sedentary endoparasitic nematodes, such as root-knot (*Meloidogyne* spp.) and cyst (*Heterodera* and *Globodera* spp.) nematodes (Roberts, 1992; Starr & Roberts, 2004). In such varieties, the nematodes can penetrate the plant roots, but they cannot establish their feeding cells there, i.e. the giant cells of the root-knot nematodes and the syncytia of the cyst nematodes. Thus the nematodes cannot develop to the adult stage. Crops for which resistant varieties are available against migratory nematodes, such as lesion (*Pratylenchus*) and stem and bulb (*Ditylenchus*) nematodes, are very rare, although many sources of resistance to migratory nematodes are known (Peng & Moens, 2003).

The most famous and popular nematode resistance in vegetable crops is that in tomato cultivars to root-knot nematodes. The resistance gene (*Mi*) originated from a wild tomato (*Lycopersicon peruvianum*) in Peru, and it was introduced into *Lycopersicon esculentum* more than 60 years ago. Many of the tomato varieties developed in the last decade carry this gene. Tomato varieties possessing this gene are resistant to the three major species of root-knot nematodes (*Meloidogyne incognita, M. javanica* and *M. arenaria*), but are susceptible to *M. hapla*, which is the dominant root-knot nematode in cool regions (Williamson, 1998). However, intensive use of nematode-resistant tomato varieties has led to the development of resistance-breaking strains of root-knot nematodes in several countries (Kaloshian *et al.*, 1996; Anwar & McKenry, 2002; Tzortzakakis *et al.*, 2005). Another problem with the resistance gene in tomato is that the conferred resistance is temperature-dependent: soil temperatures above 28°C often break the resistance (Williamson, 1998). Pepper is another important crop in which several resistance genes have been identified. Some are stable at high temperatures and resistant to the three major root-knot nematodes (Djian-Caporalino *et al.*, 2007). The main target root-knot nematode for pepper resistance is *M. incognita.* Several varieties resistant to this nematode have been released (Fery *et al.*, 1998); however, the resistance varies depending on the region, probably because of differences in nematode races (of *M. incognita* and *M. arenaria*) in each region (Oka *et al.*, 2004). Pepper varieties resistant to root-knot nematodes are also suggested for use in crop rotations to reduce the nematodes' population densities in the soil (Thies & Fery, 2002). Pepper varieties are generally highly resistant to *M. javanica* (Oka *et al.*, 2004), so pepper can be used for crop rotation in *M. javanica*-infested fields. Other crops for which resistant varieties to root-knot nematodes are available include cotton, tobacco, peanuts and soybean, among others (Starr & Roberts, 2004).

Other important resistances in crops are those to cyst nematodes. Cyst nematodes are the most important pathogens in potato, soybean and sugar beet. Because these crops are grown in open fields on a very large scale and their crop value is relatively low, chemical control is not always cost-effective or feasible. Potato cultivars resistant to the potato cyst nematode *Globodera rostochiensis* were developed and released in the 1980s; however, those resistant to *G. pallida* were only released later. A high percentage of the potato cultivars grown in countries with potato cyst nematode problems are resistant to the nematodes (Starr & Roberts, 2004). However, nematode pathotypes complicate the development of nematode-resistant cultivars. Cultivars with more durable resistance to the different pathotypes have yet to be developed (Brodie *et al.*, 2000). Soybean varieties resistant to *Heterodera glycines* were released in the 1970s; however, many races of this nematode exist, based on host status of several soybean cultivars and lines (Riggs & Schmitt, 1988). Selection of effective cultivars for the nematode(s) in a field requires identification of the nematode pathotype(s). Several cultivars resistant to cyst nematodes are also available in wheat and barley to *H. avenae*, in clover to *H. trifolii*, and in flue-cured tobacco to *G. tabacum,* to name a few (Starr *et al.*, 2002). Agronomically-desirable nematode-resistant sugar beet cultivars to *H. schachtii* are unavailable.

Unfortunately, no cultivars resistant to root-knot nematodes are available in the main cucurbitaceous crops, including cucumber, melon, watermelon, zucchini and pumpkin. Use of nematode-resistant varieties for long periods may cause the establishment of resistance-breaking strains of nematodes, similar to the resistance-breaking *Meloidogyne* spp. that occurred in tomato, thereby replacing the dominant nematode species on certain crops with other nematode species. For example, long-term use of potato varieties resistant to *G. rostochiensis* caused *G. pallida*, which had been a minor species of potato cyst nematode, to become dominant in fields in the UK (Minnis *et al.*, 2002).

4. Grafted seedlings

Use of nematode-resistant rootstocks is very common for stone-fruit trees. Several nematode-resistant rootstocks are available, mainly for *Meloidogyne* and *Pratylenchus* spp. (Fernandez *et al.*, 1994; Alcaniz *et al.*, 1996; Rubio-Cabetas *et al.*, 2000). Grafting technology in fruit-bearing vegetable crops was developed and used mainly in Japan and Korea for about 90 years (Oda, 1999) before the technique spread to other countries. Grafted seedlings are used to improve plant growth by overcoming abiotic and biotic stresses, such as low/high temperature, salinity, and a variety of soilborne diseases. Use of grafted seedlings is most popular in cucurbitaceous and solanaceous crops, such as melon, watermelon, cucumber, tomato and eggplant. This technique enables the immediate "introduction" of resistance/tolerance to soilborne diseases in commercial varieties. Intergeneric grafting is also used for vegetables, such as cucumber, watermelon and melon. The disadvantages of using grafted seedlings are the higher cost per seedling and possible incompatibility between rootstock and scion, which may cause yield reductions, even under optimum field conditions. The use of grafted seedlings became accepted in other countries after the phase-out of methyl bromide and the restrictions on the use of other nematicides and soil fumigants. Resistant rootstocks for vegetables to root-knot nematodes are available in tomato, eggplant and pepper, and are a very important tool for reducing nematode damage in organic farms. Nematode-resistant tomato rootstocks have the same problems as nematode-resistant varieties (i.e., temperature dependence of the resistance and occurrence of resistance-breaking races), because they are likely to have the same resistant gene. However, grafted tomato plants are usually more vigorous and tolerant than non-grafted susceptible tomato varieties, even when they are infected with root-knot nematodes. The rootstocks probably carry other genes which contribute to the plant's tolerance.

In eggplants, intraspecific grafting is usually applied, such as *Solanum integrifolium* and *S. torvum*, mainly to avoid the verticillium and fusarium wilt caused by *V. dahliae* and *Fusarium oxysporum* f. sp. *melongenae*, respectively. The rootstock, *S. torvum*, is also known to have high resistance to root-knot nematodes, and is used commercially (Ali *et al.*, 1992; Oda, 1999; Ioannou, 2001). In pepper, use of grafted seedlings is not as popular as with tomato and eggplant, because the grafted seedlings barely improve plant growth or increase yield, even when abiotic and biotic stresses are absent in the field. Compatibility problems between rootstocks and scions are still obvious in many cases (Oka *et al.*, 2004), and further development of better rootstock varieties is needed. However, pepper plants grafted onto

Fig. 6. Use of grafted pepper seedlings against *Meloidogyne incognita*. A: Growth of pepper plants grafted on susceptible (front) and resistant (back) rootstocks in *M. incognita*-infested soil. B: Roots of susceptible (left) and resistant (right) rootstocks grown in the infested soil.

nematode-resistant rootstocks resulted in higher yield in an *M. incognita*-infested field than non-grafted plants or plants grafted on a nematode-susceptible rootstock (Fig. 6) (Oka *et al.*, 2004).

Grafting is well accepted in cucurbitaceous crops (melon, watermelon and cucumber) in several counties. Improved growth and resistance to some fungal diseases (mainly fusarium wilt) are acquired in grafted cucurbitaceous crops; unfortunately, rootstocks resistant to root-knot nematodes are not currently available. Cucurbits are highly susceptible to root-knot nematodes, and nematode control is essential for these crops. Plant species used as rootstocks for cucurbitaceous crops are: *Cucumis metuliferus* and *C. anguria* for melon, *Sicyos angulantus* and *Cucurbita ficifolia* for cucumber, and *C. ficifolia*, *Cucurbita moschata*, *Lagenaria siceraria* and *Benincasa hispida* for watermelon. Grafted cucurbitaceous plants appear to be more tolerant than non-grafted plants to nematode infestation. Cantaloupe melons grafted on *C. moschata* were more tolerant, but not resistant, to the root-knot nematode *M. incognita* than non-grafted plants (Sigüenza *et al.*, 2005). Grafting susceptible melons onto *C. metuliferus* rootstocks reduced the levels of root galling, prevented losses in shoot weight losses, and resulted in significantly lower nematode levels at harvest (Sigüenza *et al.*, 2005). Since several reports of some degrees of resistance to root-knot nematodes in Cucurbitaceae have been published (Krishnaveni & Subramanian, 2005), future development of resistant rootstocks is expected.

5. Crop rotation

Crop rotation is a traditional cultural practice to keep pest and pathogen populations below economic thresholds and maintain soil fertility. Crop rotation with non-host plants to certain nematode species can reduce nematode populations in the field (Table 3). A non-host plant may be a cash crop or non-cash crop (a cover crop), which may serve as green manure via incorporation into the soil or for other purposes, for example increasing nitrogen in the

Table 3. Cover, rotation or green manure crops used for nematode suppression*

Nematode	Cover, rotation or green manure crop	Main crop	Reference
Belonolaimus longicaudatus	American joint vetch (*Aeschynomene americana*), hairy indigo (*Indigofera hirsute*)	Cabbage ,bean	Rhoades, 1976; Rhoades & Forbes, 1986
Ditylenchus destructor	Buckwheat, carrot, lupine, small grains, vetch	Potato	Kiryanova & Krall, 1971
Globodera pallida and *G. rostochiensis*	Non-solanaceous crops	Potato	Hijink, 1972; Trudgill *et al.*, 1987; Tiilikkala, 1991
Heterodera avenae	Maize, rye, vetch	Wheat, barley, oat	Bachthaler & Hien, 1976; Romero *et al.*, 1991
Heterodera glycines	American joint vetch, bahia grass (*Paspalum notatum*), cotton, hairy indigo, maize, sorghum, velvet bean (*Mucuna pruriens*)	Soybean	Francl & Dropkin, 1986; Rodriguez-Kábana *et al.*, 1988b, 1990a, 1991c; Weaver *et al.*, 1988, 1998
Heterodera schachtii	Potato, cereals, bean, maize, navy bean, lucerne, onion, pea, snap bean (*Phaseolus vulgaris*), resistant mustard (*Sinapus alba*), resistant oil radish (*Raphanus sativus*), wheat	Sugar beet	Golden & Shafer, 1959; Steele & Price, 1965; Fichtner *et al.*, 1984; Cooke, 1985; Tacconi & Regazzi, 1990; Müller, 1991; Koch & Gray, 1997
Hoplolaimus columbus	Bahia grass, pepper, sweet potato, tall fescue, tomato	Soybean	Lewis & Fassuliotis, 1982
Meloidogyne arenaria	American joint vetch, bahia grass, cotton, *Crotalaria spectabilis*, hairy indigo, maize, partridge pea (*Cassia fasciculate*), sesame, sorghum, sorghum-Sudan grass, velvet bean	Groundnut	Rodriguez-Kábana & Ivey, 1986; Rodriguez-Kábana *et al.*, 1988a, 1990b, 1991a,b, 1994; McSorley *et al.*, 1994

Table 3. Cover, rotation or green manure crops used for nematode suppression (continued)

Nematode	Cover, rotation or green manure crop	Main crop	Reference
Meloidogyne chitwoodi	Barley, lucerne, oat, rape plant, sorghum-Sudan grass, Sudan grass, wheat	Potato	Viglierchio, 1987; Mojtahedi *et al.*, 1993
Meloidogyne graminicola	*Digitaria adscendens*, guzitil (*Guizotia abyssinica*), jute, millet (*Eleusine indica*), mustard, okra, pepper	Rice	Roy, 1979, 1987; Rahman, 1990
Meloidogyne hapla	Barley, *Hesperis matronalis*, *Lolium multiflorum*, maize, oat, radish, rye, spinach, Sudan grass, *Tagetes minuta*, wheat	Several	Ngundo, 1979; Bélair, 1992; Lamers & Roosjen, 1993; Widmer & Abawi, 2000, 2002
Meloidogyne incognita	Bahia grass, barley, Bermuda grass (*Cynodon dactylon*), castor, cotton, *Crotalaria juncea*, *C. resuta*, *C. spectabilis*, groundnuts, lucerne, *Macroptilium atropurpureum*, maize, mustard, pangola grass (*Digitaria decumbens*), *Panicum maximum*, rye , sorghum, sesame, *Tagetes patula*, *T. erecta*, wheat	Several	Ayala, 1971; Bergeson, 1971; Ducusin & Davide, 1972; Murphy *et al.*, 1974; Netscher, 1974; Carter & Nieto, 1975; Di Muro, 1975; Johnson *et al.*, 1975; Belcher & Hussey, 1977; Motsinger *et al.*, 1977; Shagalina, 1979; Huang *et al.*, 1980; Sharma *et al.*, 1980; Carneiro & Carneiro, 1982; Atu, 1984; Ichinohe, 1984; Araki *et al.*, 1991; Prasad *et al.*, 1992; McSorley & Dickson, 1995; Ploeg, 2000, 2002
Meloidogyne javanica	Cassava, cotton, *Chloris gayana*, groundnut, *Crotalaria juncea*, *C. ochroleuca*, *Erabrostis curvula*, maize, *Panicum maximum*, sesame, *Tagetes erecta*, *T. minuta*, *T. patula*, wheat	Several	Daulton, 1963; Whitehead, 1965; Birat, 1970; Mitchell *et al.*, 1971; Saka, 1982; Abid & Maqbool, 1990; Toida *et al.*, 1991; Madulu *et al.*, 1994
Nacobbus aberrans	Cotton, groundnuts, melon, okra, sesame, soybean, squash, sunflower, watermelon	Several	Bridge, 1983

Table 3. Cover, rotation or green manure crops used for nematode suppression (continued)*

Nematode	Cover, rotation or green manure crop	Main crop	Reference
Paratrichodorus minor	Asparagus, *Crotalaria juncea,*	Several	Rhode & Jenkins, 1957
Pratylenchus brachyurus	Groundnut	Several	Good *et al.*, 1954
Pratylenchus loosi	Guatemala grass (*Tripsacum laxum*), mana (*Cymbopogon confertifolius*), weeping love grass (*Eragrostis curvula*)	Tea	Sivapalan, 1972; Gnanapragasam, 1981
Pratylenchus penetrans	Pearl millet, Saia oat (*Avena strigosa*), oat, *Tagetes erecta, T. patula*	Several	Oostenbrink *et al.*, 1957; Caubel *et al.*, 1978, 1981; Chikaoka, 1983; Reynolds *et al.*, 2000; Alexander & Waldenmaier, 2002; Bélair *et al.*, 2004; LaMondia, 2006
Pratylenchus thornei	Beet, *Cannabis sativa*, carrot, coriander, *Nasturtium fantarum*, parsley, radish, spinach, sugarbeet, turnip	Chickpea	Greco *et al.*, 1988
Pratylenchus zeae	Cowpea, *Crotalaria juncea*, mung bean	Rice, sugarcane	Aung & Prot, 1990; Sundararaj & Mehta, 1990
Radopholus similis	*Brachiaria decumbens*, *Digitaria milanjiana, Macroptilium atropurpureum,* sorghum	Banana	Ternisien, 1989; Ternisien & Melin, 1989
Rotylenchulus reniformis	*Brassica juncea*, *B. campestris*, *B. napus, Crotalaria juncea*, maize, pangola grass, sorghum, sugarcane, *Tagetes patula*	Pineapple, cotton, sweet potato	Ayala, 1971; Nakasono, 1973; Braithwaite, 1974; Thames & Heald, 1974; Wang *et al.*, 2001
Xiphinema americanum	Barley, *Crotalaria,* millet, rye, oilseed rape, Sudan grass, wheat	Chicory, fruit frees, etc.	Brodie *et al.*, 1970; Boldyrev & Borzykh, 1983; Halbrendt & Jing, 1994
Xiphinema index	Barley, *Brassica juncea, Ruta graveolens, Thymus vulgaris*	Grapevine	Aballay *et al.*, 2004

* Based mainly on Whitehead, 1997

soil by nitrogen-fixing plants, weed control and prevention of soil erosion. Great care should be paid to selecting the crops (or even the varieties), as well as the cover crops, for the rotation, because some nematode species have very wide host ranges while others have very narrow ones. Target nematode(s) in the field should be identified before selecting plants. The nematode control efficacy achieved by rotating crops may be highly cultivar- and nematode-specific. Even in a nematode species, the host range varies among nematode races or pathotypes. For example, variations exist in host range (plant species and crop cultivars) of some root-knot and cyst nematodes, such as *M. incognita, M. arenaria* and *H. glycines, H. avenae* and *G. rostochiensis*. Susceptibility to other diseases and insects should also be considered before a crop's introduction into the rotation. Generally, cyst nematodes (*Heterodera* and *Globodera* spp.) have narrower host rages than root-knot (*Meloidogyne* spp.) and lesion (*Pratylenchus* spp.) nematodes. More than 3,000 plants are thought to be attacked by the root-knot nematode *M. incognita*. Potato, soybean, sugar beet and cereal cyst nematodes can be controlled by crop rotations with non-hosts that last several years, since these nematodes survive in the soil for years as cysts (dead females containing eggs). Generally, 4 to 6 years of a crop cycle with non-host plants prevents economic damage caused by the cyst nematodes. Among the cyst nematodes, the cereal cyst nematodes (*H. avenae* group) are easier to control by crop rotation because the nematode eggs hatch independently of the hatching factors that are present in the host root's exudates. A single fallow year reduced the population of *H. avenae* by 56 to 75%, and wheat could be grown once in 3 years (Meagher & Rooney, 1966). In contrast, potato cyst nematodes, the sugar beet cyst nematode and the soybean cyst nematode hatch better in the presence of hatching factors from roots of host or related plants, so the reduction rate of these nematodes is lower in each year of crop rotation as compared with the cereal cyst nematodes. The number of unhatched eggs of *G. pallida* and *G. rostochiensis* was reported to decline at an annual rate of 12.8 to 40.5% in barley plots (Whitehead, 1995). The population density of the potato cyst nematode, *G. rostochiensis*, was kept under the economic threshold by cropping a susceptible cultivar once in 4 or 5 years (Tiilikkala, 1991). It is recommended that beets be planted every 4 to 6 years with no crucifers planted in the rotation to prevent damage from *H. schachtii* (Westphal & Becker, 2001). A 5-year rotation of non-host crops was needed to lower *H. glycines* population densities to below damaging levels (Chen *et al.*, 2001b). Rotation periods could be shortened for these cyst nematodes if resistant varieties were used.

For nematodes that need plant stimuli for hatching, such as some cyst nematodes, susceptible or resistant crops which contain hatching factors in their root exudates may be used as trap crops. Such crops are grown for a short period to enhance nematode hatching, the nematode juveniles are allowed to invade the plant roots, and then the susceptible crop is killed by incorporation into the soil before the females lying lay eggs. If resistant crops or varieties are used, they can be grown for longer periods. Resistant cruciferous crops can be used as trap crops for the control of *H. schachtii* (Cooke, 1985; Koch & Gray, 1997). Trap crops have been used for *H. glycines* (Chen *et al.*, 2001a), *G. rostochiensis* (Whitehead, 1977; Scholte, 2000) and for root-knot nematodes using Brassicaceae plants, such as oilseed radish (*Raphanus sativus*) or arugula (*Eruca sativa*), although most root-knot nematodes do not need plant stimuli for hatching (Melakeberhan *et. al.*, 2006, 2008).

For nematodes with broad host ranges, mainly root-knot and lesion nematodes, non-cash cover crops are used in the crop rotation to lower nematode populations in the field (Table 3). Potential cover crops in crop rotations for nematode control include American jointvetch

(*Aeschynomene americana*), bahiagrass (*Paspalum notatum*), castor (*Ricinus communis*), rye (*Secale cereale*), sesame (*Sesamum indicum*), sorghum (*Sorghum bicolor*), sorghum-sudangrass hybrids (*Sorghum bicolor* × *S. bicolor* var. *sudanense*), sunnhemp (*Crotalaria juncea*), marigold (*Tagetes* spp.), velvet beans (*Mucuna pruriens*), vetch (*Vicia sativa*) and more (Rodríguez-Kábana *et al.*, 1991c, 1992; Johnson *et al.*, 1992; McSorley *et al.*, 1994; Viaene & Abawi, 1998; Weaver *et al.*, 1998; McSorley, 1999; Crow *et al.*, 2001; Kimpinski *et al.*, 2000; Widmer & Abawi, 2002). *Tagetes* spp. and *Crotalaria* spp. have received special attention because of their resistance/non-host status to several root-knot and lesion nematodes and the reniform nematode *Rotylenchulus reniformis* (Wang *et al.*, 2001, 2002a, 2004; Alexander & Waldenmaier, 2002; Ploeg, 2002), and the nitrogen-fixing ability of *Crotalaria* spp. *Tagetes* spp., mainly *T. erecta* and *T. patula*, are known for their nematicidal effect on lesion and root-knot nematodes. Nematicidal compounds have been isolated from the plants' roots (Gommers & Bakker, 1988), but their involvement in nematode control in the soil rather than in plant roots has not been confirmed. *Tagetes* varieties suitable for nematode control have been developed, such as *T. patula* 'Nema-Gone' and the *Tagetes* hybrid 'Polynema' (Ploeg, 2002). In the Kanagawa prefecture of Japan, growing *Tagetes* spp. before growing Japanese radish is a well-accepted practice to control the lesion nematode *Pratylenchus penetrans* (Chikaoka, 1983), and other crops are grown in other countries for the same purpose (Kimpinski *et al.*, 2000). Cultivars of *T. patula* grown before flue-cured tobacco have been shown to be a cost-effective alternative to chemical soil fumigation for control of *P. penetrans* (Reynolds *et al.*, 2000). *T. erecta* was also effective in reducing the same lesion nematode before cropping potato or tomato (Alexander & Waldenmaier, 2002). Yields of tomato and melon were increased by pre-cropping *T. patula* and the *Tagetes* hybrid 'Polynema' due to the reduction in field *M. incognita* infestations (Ploeg, 2002). Several *Crotalaria* spp. are poor hosts to *Meloidogyne* spp., *Rotylenchulus reniformis*, *Radopholus similis*, *Belonolaimus longicaudatus* and *H. glycines*, and their use as a rotation crop has been shown to be effective in the control of several other nematode species (Wang *et al.*, 2002a). The nematode control effect of sunnhemp has been tested intensively on *R. reniformis* on pineapple in Hawaii (Wang *et al.*, 2001, 2002b, 2003). Suppression of root-knot nematodes by *Crotalaria* spp. has also been confirmed in several studies (Wang *et al.*, 2002a). Nematicidal compounds, such as monocrotaline and pyrrolizidine alkaloids, have been found in *Crotalaria* spp., but their involvement in nematode reduction is not clear (Fassuliotis & Skucas, 1969; Jourand *et al.*, 2004).

Sudangrass and sorghum-sudangrass hybrids are known to produce a chemical, glycoside dhurrin, which degrades into nematicidal hydrogen cyanide (Widmer & Abawi, 2000, 2002). A correlation between nematode reduction efficacy and hydrogen cyanide contents has been indicated in sorghum-sudangrass hybrids (Widmer & Abawi, 2002). These hybrids were used as a cover crop in a potato production system in northeast Florida against root-knot nematodes, but the plant proved to be a good host for the sting nematode *Belonolaimus* spp. (Weingartner & Shumaker, 1983), or nematode suppression was species-specific (MacGuidwin & Layne, 1995). Rye is a potential rotation crop to reduce root-knot nematodes. The incorporation of rye (*Secale cereal*) before planting cotton reduced damage caused by *M. incognita* (McSorley & Dickson, 1995). Hydroxamic acids are suggested to play a role in nematode control by rye incorporation (Zasada *et al.*, 2005). High mortality of *M. incognita* juveniles and mixed-stages of *Xiphinema americanum* was obtained by exposure to some hydroxamic acids at a concentration of 90 μg/ml. Concentrations of nematicidal compounds in a cover crop may serve as a potential indication for the selection of varieties for nematode control (Zasada *et al.*, 2007). Plants that produce nematicidal chemicals are termed antagonistic to nematodes. Use of such plants as cover crops in a rotation may be more effective in nematode control than cropping non-host plants without nematicidal compounds (Halbrendt, 1996). The rotation must be accompanied by complete eradication of all susceptible weeds or volunteer plants, because nematodes may proliferate on such plants during the rotation, maintaining their population at damaging levels.

Economic loss during the growing period of non-cash crops is the main disadvantage of this control method, especially in intensive organic farms in greenhouses, although incorporation of cover crops into the soil as green manure may fertilize and enrich the diversity of microorganisms in the soil, and green manure also occasionally controls nematodes. Intercropping is another potential way to reduce nematode population density by planting non-host plants or antagonistic plants together with a main crop. In this method, the economic loss caused by growing non-crops for a season(s) is reduced. However, the intercrop also needs to be chosen very carefully. For example, sunnhemp or rapeseed intercropped with pineapple in Hawaii reduced the population of *R. reniformis*, but the rapeseed increased that of *M. javanica*, and the sunnhemp shaded the pineapple plants (Wang *et al.*, 2003).

6. Green manures and biofumigation

A cover crop used in a crop rotation can be incorporated into the soil as green manure to increase the soil organic matter content and fertility, and reduce the nematode population, via an increase in nematode-antagonistic microorganisms, the absence of host plants or nematicidal compounds released from the degrading green manure (Stirling, 1991; Sikora, 1992; D'Addabbo, 1995; Mitkowski & Abawi, 2002). Plants containing nematicidal compounds or their precursors are preferred as green manure for the suppression of nematode populations. As mentioned in the previous section, sudangrass and sorghum-sudangrass hybrids, which contain cyanide precursors, have been used as green manure for root-knot nematodes (Mojtahedi *et al.*, 1993; Widmer & Abawi, 2002). Cultivars of pearl millet and other rotation crops have been tested for control of *P. penetrans* on flue-cured tobacco (Bélair *et al.*, 2004). CFPM 101, a pearl millet cultivar containing greater concentrations and amounts of sulfur-containing compounds, showed promising results when it was grown and then incorporated in the soil before tobacco planting. Saia oat (*Avena strigosa*) and oat (*Avena sativa*) used as rotation and green manure crops reduced *P. penetrans* on potato after 2 years of rotation (LaMondia, 2006).

Other green manure plants containing nematicidal compounds or precursors are *Tagetes* spp., *Crotalaria* spp. and Brassicaceae plants. Many nematicidal compounds have been isolated from Asteraceae plants (Gommers & Bakker, 1988). Marigolds produce polythienyls and thiophenes, which are toxic to nematodes (Gommers & Bakker, 1988). The most famous nematicidal compound from this plant family is α-terthienyl, found in the roots of *Tagetes* spp. (Uhlenbroek & Bijloo, 1958). However, involvement of this compound in the soil rather than inside the plant roots is not known. Incorporation of *Tagetes* spp. foliage or extracts has also been reported to reduce nematode populations. Pre-cropping of *T. patula* for 8 weeks in *M. incognita*-infested soil reduced galling on tomato roots and the final nematode population (Ploeg, 2000). Nematicidal activity was confirmed in the plant foliage of *T. erecta* (Natarajan *et al.*, 2006), and several volatile and non-volatile nematicidal compounds were isolated from flowers of *T. erecta* (Debprasad *et al.*, 2000). However, involvement of these nematicidal compounds in nematode suppression via use of the foliage as green manure is not clear. Black-eyed Susan (*Rudbeckia hirta*) is a non-host plant of some root-knot nematodes, and suppresses populations of other plant-parasitic nematodes (Thetford & Kinloch, 1999). Thiarubrine C, a nematicidal compound, was isolated from the roots of that plant (Sànchez de Viala *et al.*, 1998). Field chrysanthemum (*Chrysanthemum corona*), incorporated into soil fresh or dried, reduced the population of the root-knot nematode *M. javanica* in the soil (Bar-Eyal *et al.*, 2006). An aqueous extract of *C. coronarium* was nematostatic to juveniles, and inhibited their hatching from eggs. The leaf powder of *Inula viscosa*, which is grown in the Mediterranean basin, was found to control *M. javanica* and other plant-parasitic nematodes when it was incorporated into the soil (Oka *et al.*, 2001). Nematicidal sesquiterpenic acids were isolated from that plant's leaves.

Plants belonging to the family Brassicaceae have been widely used as green manures to control soilborne pathogens, including nematodes (Kirkegaard & Sarwar, 1998; Curto *et al.*, 2005; Matthiessen & Kirkegaard, 2006). These plants are known to contain glucosinolates, which are broken down into isothiocyanates (ITCs) after their incorporation into the soil. ITCs are compounds similar to the active compound (methyl ITC) generated from the commercial soil fumigant metham sodium. Methyl ITCs have nematicidal, herbicidal and fungicidal activities. ITCs from Brassicaceae were found to be nematicidal to *M. incognita* juveniles at low concentrations in *in-vitro* tests (LD_{50} of 11-34 μM) (Lazzeri *et al.*, 2004a). This control method for soilborne diseases, which utilizes volatile pesticidal compounds from degrading plant residues incorporated in the soil is called "biofumigation" (Matthiessen & Kirkegaard, 2006). Rapeseed and mustard are the major crops used for biofumigation (Mojtahedi *et al.*, 1991; Ploeg & Stapleton, 2001, Lazzeri *et al.*, 2004b; Zasada & Ferris, 2004). Variations in glucosinolate profiles and their concentrations exist among Brassicaceae crops and their developmental stages (Bellostas *et al.*, 2007). Rocket or arugula (*Eruca sativa*) was also found to be a potential biofumigant plant due to its high resistance to root-knot nematodes and the presence of glucosinolates (Lazzeri *et al.*, 2004a; Curto *et al.*, 2005). Plant residues of cruciferous cash crops, such as broccoli, cabbage and cauliflower, can be incorporated after harvest into the soil for biofumigation (Gamliel & Stapleton, 1993; Stapleton & Duncan, 1998; Ploeg & Stapleton, 2001). Although some studies have shown that glucosinolates are involved in the nematicidal activity of these plants (Lazzeri *et al.*, 1993; Potter *et al.*, 1999), this is not always clear (McLeod & Steel, 1999). Glucosinolates are thought to play an important role in nematode suppression, although other chemicals, such as sulfur-containing compounds, as well as biological factors, may also be involved. Again, before selecting a plant for biofumigation, the host range of the target nematode(s) in the field needs to be checked. Non-host or resistant varieties of biofumigant plants are highly preferred. A study in Australia showed that *M. javanica* reproduces on *Brassica juncea* and *B. napus* at rates which increase the nematode population in the soil during their growth period in hot seasons; however, nematode proliferation could be prevented if they were grown in the winter, under low temperatures (Stirling & Stirling, 2003). To obtain the maximum control efficacy from biofumigation, several factors must be considered in addition to host status to nematodes, including: the optimum plant growth stage for soil incorporation, probably when the glucosinolate content reaches maximum levels, the biomass of the plant, and adequate soil moisture for plant degradation and gas diffusion into the soil while preventing rapid gas escape from the soil to the atmosphere. The soil surface can be tarped with a plastic film after plant incorporation into the soil to enhance fumigation efficacy of ITCs, if applicable and cost-effective. If tarping is used in hot seasons, the soil temperature is also elevated and further improvement of control efficacy is obtained by the soil solarization effect (see also *Soil solarization* section in this chapter) (Gamliel & Stapleton, 1993; Stapleton & Duncan, 1998; Ploeg & Stapleton, 2001; Guerrero *et al.*, 2005). Use of dried pellets of glucosinolate-containing plants is another technique to gain a biofumigation effect without growing biofumigation plants in the field (Lazzeri *et al.*, 2004c).

Other plants used for green manures include alfalfa, clover, lupine and buckwheat: although these are not generally used for the control of nematodes with broad host ranges, such as *Pratylenchus* and *Meloidogyne* spp., they can be used for nematodes with very specific host plants.

7. Organic amendments

Soil application of organic amendments, mainly compost, is essential in organic farming in order to maintain soil fertility and a soil structure that is suitable for crop production. The effect of organic amendments on nematodes has been known for decades (Lindford *et al.*, 1938), and many types of organic amendments have been used for nematode control (Rodríguez-Kábana, 1986; Stirling, 1991; D'Addabbo, 1995; Widmer *et al.*, 2002). Two main mechanisms are involved in nematode control by organic amendments: 1) release of nematicidal compounds from organic amendments into the soil; 2) amendment-induced population build-up of microorganisms that are antagonistic to nematodes. The latter is mentioned in the *Biological control* section of this chapter.

Amendments with nematicidal compounds or precursors: One type of organic amendment used for nematode control is plant material containing nematicidal compounds or their precursors. Such compounds may be released into the soil over a short period of time. Green manure crops containing nematicidal compounds and precursors were mentioned in the previous section. The best known nematicidal amendments are several preparations from seeds or leaves of the neem tree (*Azadirachta indica*) (Mojumdar, 1995). Several commercial preparations from this tree are available for insect, nematode and fungal disease control (Mojumdar, 1995; Raguraman *et al.*, 2004; Suresh *et al.*, 2004). It is mainly the kernel of the plant seeds that is used for pest control, but leaves are also reported to have nematicidal activity. Many pesticidal compounds, mainly limonoids, have been identified from this plant. Among them, azadirachtin is the major compound for pest control. Oilcake of neem is usually one of the most effective and feasible amendments for nematode control and soil fertility (Akhtar, 1998). However, nematode control efficacy is not always satisfactory, probably due to the low concentration of nematicidal compounds in the oilcake. Sesame oilcake may also contain nematicidal compound(s) because it is very effective at controlling *M. incognita* (Radwan *et al.*, 2009), and natural "nematicides" based on sesame oil and stalk are commercially available. As already mentioned, Brassicaceae plants are used for biofumigation by incorporating plants grown as a cover crop. To eliminate the growing period required for cover crops in intensive farming systems, dry foliages of biofumigant plants have been tested against soilborne diseases (Lazzeri *et al.*, 2004c). The dried plants had sufficient residual content of glucosinolate and myrosinase activity after water addition, and showed fungitoxic activity. A pellet based on *Brassica carinata* defatted seed meal showed nematicidal activity after its incorporation into soil infected with *M. incognita* (Curto *et al.*, 2008).

Amendments that release nematicidal compounds during decomposition: A variety of organic amendments that release nematicidal compounds during their decomposition in the soil have been used for nematode control (Rodríguez-Kábana, 1986; Stirling, 1991; D'Addabbo, 1995). The most popular and feasible amendment for nematode control is animal manure (Kaplan & Noe, 1993; Lazarovits et al., 2000; Riegel & Noe, 2000; Oka & Yermiyahu, 2002; Oka et al., 2007a), regardless of whether it has been composted or not. Other organic materials used for this purpose are oilcakes, cottonseed meal, crab shells, feather meal, bone meal, blood meal, horn meal, fungal mycelium, etc. (Walker, 1971; Rodríguez-Kábana, 1986; Stirling, 1991; D'Addabbo, 1995). These materials have high nitrogen-to-carbon ratios. When these organic materials are degraded by soil microorganisms, ammonia (NH_3) is released. Ammonia is known to be involved in the decline of plant-parasitic nematodes and plant-pathogenic fungi following application of organic amendments with high nitrogen-to-carbon ratios (Rodríguez-Kábana, 1986; Rodríguez-Kábana et al., 1987; Spiegel et al., 1987; Oka et al., 1993; Tenuta & Lazarovits, 2002). These amendments should be applied long before seeding or transplanting in order to avoid phytotoxicity caused by ammonia and other compounds; alternatively, a supplemental carbon source can be used to reduce the phytotoxicity (Rodríguez-Kábana et al., 1987). The mode of action of ammonia in pathogen control is not clear, but several mechanisms are thought to be involved: membrane disruption (Rush & Lyda, 1982), elimination of proton gradients across membranes (Docherty & Snider, 1991) and exhaustion of the chemical energy of cells removing cytosolic ammonia across a concentration gradient (Britto et al., 2001). Ammonium (NH_4^+), the ionized form of ammonia, cannot cross the cell membrane, and is not nematicidal. The use of chemical ammonia (anhydrous or aqueous ammonia) itself has not been accepted commercially for control of nematodes and fungi, although its potential effectiveness has been demonstrated (Rodríguez-Kábana et al., 1981; Oka & Pivonia, 2002).

A limitation to the use of ammonia-releasing organic amendments for nematode control is that the large amounts required for satisfactory control are often phytotoxic or simply not cost-effective (Rodríguez-Kábana et al., 1987; Stirling, 1991). Generally, high application rates of organic amendments (>1,600 kg N/ha or amendments at 20 t/ha or more) are required for nematode control, although nematode control efficacy is highly dependent on soil conditions, and it can be enhanced by manipulating the soil environment. The application rates required for efficient nematode control are highly dependent on soil pH, since the equilibrium of ammonia in the liquid phase of the soil with the non-nematicidal ammonium is pH-dependent, as formulated by the Henderson-Hasselbalch equation ($\log [NH_{3(g)} / NH_4{}^+{}_{(aq)}] = pH - 9.3$) (Duplessis & Kroontje, 1964). In pH 9.3 soil, only 50% of the applied ammonia remains as ammonia, and the other half converts to non-nematicidal ammonium; if the pH is elevated to about 11, ammonia becomes dominant, and reaches nearly 100% of the total amount applied. Another limiting factor for the use of ammonia-releasing amendments is the instability of ammonia in soil. Ammonia is readily oxidized to nitrite (and then to nitrate) by soil microorganisms. Generally, soil pH increases during the decomposition of applied amendment due to the released ammonia, but the pH drops shortly thereafter, when ammonia is oxidized to nitrite and nitrate. Two methods for enhancing the nematicidal activity of ammonia-releasing amendments appear to be feasible: 1) raising the soil pH and shifting the equilibrium toward ammonia, thus reducing the amount of amendment required for effective nematode control; 2) using nitrification inhibitors to allow ammonia concentrations to remain in the soil for extended periods. Both methods should be transient and reversible in the soil; otherwise, the ammonia will become phototoxic to the subsequent crop.

The nematicidal activity of broiler litter against *M. javanica* was enhanced by transiently increasing the soil pH via the use of alkaline amendments, such as cement kiln dust and alkaline-stabilized biosolids (Oka *et al.*, 2006b,c). Applications of broiler litter combined with these alkali amendments effectively reduced nematode infection on tomato plants grown in experimental plots. On an organic farm, application of slaked lime in combination with broiler litter slightly but significantly reduced the root-galling index of pepper plants caused by *M. incognita* (Oka *et al.*, 2007a). This technique is suitable for neutral to alkaline soils, whose pH can be relatively easily increased. The increase in soil pH is transient, returning to its original level after several weeks. The use of alkaline amendments, such as lime, must be permitted by the local organic farming authority.

Application of a chemical nitrification inhibitor (nitrapyrin) in combination with organic or chemical ammonia-releasing amendments was shown to increase the nematicidal activity of the amendments against *M. javanica* (Oka & Pivonia, 2003). Organic nitrification inhibitors may be used in organic farming systems. Some medicinal and herbal plants have been reported to have nitrification-inhibiting activity, and these may also be used to enhance the nematicidal activity of ammonia released from amendments. Extracts of spearmint (*Mentha spicata*), *Artemisia annua*, neem and karanja (*Pongamia glabra*) were tested for their nitrification-inhibiting efficacy (Sahrawat & Mukerjee, 1977; Usha & Patra, 2003). In particular, karanja and neem extracts were found to be relatively effective at inhibiting nitrification (Sahrawat & Mukerjee, 1977; Gnanavelrajah & Kumaragamae, 1999). The nematicidal activities of chicken litter and chitin, alone or in combination with neem extracts, were tested against *M. javanica*. Soil application of these organic amendments or the neem extracts alone reduced the root-galling index of tomato plants only slightly, if at all, but application of the amendments in combination with the neem extracts reduced root galling significantly (Oka *et al.*, 2007b). Soil analysis indicated that the neem extract inhibited nitrification of the ammonia released from the amendments and extended the persistence of the ammonia concentrations in the soil. The extended exposure of nematodes to ammonia as a result of nitrification inhibition by the neem extracts appeared to be the cause of the enhanced nematicidal activity of the ammonia-releasing amendments. Since marketed neem extracts are expensive for use as a nitrification inhibitor in soil treatments, cheaper neem products, such as neem oilcake, should be tested for their nitrification-inhibition effect. Another technique to enhance the nematicidal activity of organic amendments is to combine amendment application with solarization (see *Soil solarization* section in this chapter).

Application of organic amendments with high nitrogen-to-carbon ratios is also effective in acidic soils for the control of soilborne diseases and nematodes. The compounds involved in this control are acids, rather than ammonia. As mentioned above, ammonia is released from amendments with high nitrogen-to-carbon ratios during their decomposition, and oxidized to nitrite in the first stage of nitrification by soil microorganisms, mainly *Nitrosomonas* bacteria. In acidic environments, non-nematicidal ammonium is dominant, the nitrification process is not as fast as in alkaline soil, and nitrite is occasionally accumulated. Nitrite (NO_2^-) in the soil water phase is in equilibrium with nitrous acid [HNO_2], the unionized form of nitrite. Nitrous acid is known to be highly toxic to fungi and nematodes (Tsao & Oster, 1981; Oka *et al.*, 1993; Tenuta & Lazarovits, 2002): it is 300 to 500 times more effective at killing *Verticillium dahliae* microsclerotia than ammonia (Tenuta & Lazarovits, 2002). In contrast to ammonia, nitrous acid is dominant at low pH due to its low pKa (3.33 at 24°C). Enhanced control activity may be achieved by lowering the soil pH with some acids (Conn & Lazarovits, 2000). Swine manure is effective in at controlling the soybean cyst nematode (Xiao *et al.*, 2007), and hog manure at controlling the lesion nematode *P. penetrans* (Mahran *et al.*, 2008), which is the causal pathogen in early-dying syndrome of potato in association with the fungus *V. dahliae*. In addition to accumulated nitrous acid, short-chain fatty acids, such as acetic, propionic and butyric acids, released after application of these animal manures and from anaerobically digested slurry, are another factor in fungal and nematode control in low pH soils (Conn *et al.*, 2005; Min *et al.*, 2007; Mahran *et al.*, 2008). Butyric acid from decomposing rye and timothy residues was found to be nematicidal to *M. incognita* and *P. penetrans* at low pH (Sayre *et al.*, 1965). Higher efficacy of these short-chain fatty acids also resulted from non-ionized forms at low pH, which may be capable of crossing the pathogens' cell membranes (see also *Reductive soil disinfestation* section in this chapter).

8. Compost

Compost is a very important component in the soil, especially in organic farming systems. Composts are used to provide the soil with nutrients and organic matter and to improve the soil's structure. They are generally prepared from agroindustrial wastes, such as tree bark, animal wastes and green wastes. Some composts may also contain municipal solid wastes and sewage sludge. The composting process stabilizes the nutritional elements, converts ammoniacal nitrogen to nitrate and eliminates pathogens and weed seeds via heat generation. Another merit of using compost is a reduction in disease incidence. Some kinds of compost have been reported to suppress soilborne pathogens in the genera *Pythium*, *Phytophthora*, *Fusarium* and *Rhizoctonia* (Hoitink *et al.*, 1997). Certain microorganisms in composts are thought to play a role in plant disease control. However, the effects of composts on plant-parasitic nematodes have not been studied as thoroughly as their effects on fungal diseases. In some experiments, composts and raw sewage sludge showed some activity against root-knot nematodes (Marull *et al.*, 1997; LaMondia *et al.*, 1999) but in others, composts affected nematodes only slightly, if at all (McSorley & Gallaher, 1995, 1996). In a long-term experiment, application of compost increased total yields of potato tubers and grain yields of barley, but the impacts on *P. penetrans*, *M. hapla* and *H. trifolii* varied, usually increasing their population levels in the soil (Kimpinski *et al.*, 2003).

Cattle manure and grape marc composts were suppressive to *M. javanica* in pot and *in-vitro* experiments (Oka & Yermiyahu, 2002). No root galls were found on tomato roots grown in soils containing cattle manure compost at 10 or 25% (v/v), and very few were found on those grown in soil containing grape marc compost at 50%. Chemical analysis of the composts and leachates from the soils suggested that high EC values and high concentrations of nitrogen, especially $N\text{-}NH_4^+$, contributed to the nematode-suppressive effect of the composts. However, very high concentrations of compost, which are not feasible or cost-effective for field applications, may be needed to reduce nematode damage. Involvement of compost biological activity in nematode suppression has not been indicated. Other composts were also effective at nematode control although high percentages were needed in the soil (Raviv *et al.*, 2005). Since the nutritional status of composts and their effects on plants are extremely variable, depending on their raw materials and the composting processes, the disease-suppression mechanism may vary depending on compost types. However, abiotic and biotic factors are thought to be involved in disease suppression (Hoitink *et al.*, 1997).

9. Reductive soil disinfestation

Reductive soil disinfestation for soilborne diseases is generated by incorporating organic amendments into the soil followed by irrigation and covering the soil surface with plastic film to induce anaerobic conditions (Blok *et al.*, 2000). Application of wheat bran followed by flooding and covering the soil surface has been used to control some soilborne diseases, including fusarium and bacterial wilt of tomato (Momma *et al.*, 2006). Accumulation of acetic acid and/or butyric acid was thought to be one mechanism involved in the pathogen control by this method; however, other mechanisms have also been suggested (Momma *et al.*, 2005, 2006). Chlamydospores of *F. oxysporum* f. sp. *lycopersici*, the causal agent of fusarium wilt of tomato, were effectively killed only when the soil conditions were changed as follows: 1) oxygen was depleted by soil microorganisms after amending with wheat bran, 2) the soil surface was covered with a plastic film to establish anaerobic conditions by preventing reoxygenation from the soil surface. Successful control by this method is always correlated with a decrease in soil pH and emission of an unpleasant odor (Momma *et al.*, 2005). Although the effect of these reductive control methods on nematodes has not been tested, accumulation of organic acids, low pH and anaerobic conditions should effectively control nematodes as well. This control method may also belong to the category of *physical control methods* because the practices make the soil environment anaerobic.

Physical control methods

1. Soil solarization

Soil solarization was developed as a physical control method for soilborne diseases and weeds (Katan, 1981). Tarping moist soil with plastic film in hot seasons, usually for 1 or 2 months, increases the soil temperature to as high as ca. 50°C at a depth of 10 cm. The elevated soil temperature may directly kill soilborne pathogens and/or have a sub-lethal effect, which makes pathogens more sensitive to chemical pesticides or natural enemies (Katan, 1981). Soil solarization is generally more effective against fungal pathogens than nematodes because nematodes move from the upper soil layer to deeper ones after the host plants have been eliminated. Nematodes located at a depth of 30 to 40 cm or more are only barely affected by the heat generated by soil solarization. When plants are seeded or transplanted after the soil solarization, the nematodes move back up in the soil and infect the roots. Nevertheless, several reports have shown successful nematode control by this method (LaMondia & Brodie, 1984; Chellemi *et al.*, 1997; Giannakou *et al.*, 2007).

To enhance nematode-control efficacy of soil solarization, ammonia-releasing organic amendments have been incorporated into the soil before tarping with plastic film (Gamliel & Stapleton, 1995; Guerrero *et al.*, 2005; Oka *et al.*, 2007a). Application of cottonseed mill, soybean mill and broiler litter at application rates of 0.75 to 2.0 kg/m^2 increased control efficacy of the root-knot nematodes *M. incognita* and *M. javanica*, as compared with soil solarization or amendment application alone (Oka *et al.*, 2007a). Theoretically, high soil temperatures shift the equilibrium between ammonia and ammonium towards the former, and increase its concentration in the soil, and the plastic film prevents ammonia emission, keeping it in the soil for longer periods. Sub-lethal heating may make pathogens more sensitive to ammonia and the other toxic compounds released from the amendments. These mechanisms appear to be involved in the high efficacy of root-knot nematode control by this combination. This method also increased nematode control efficacy at the edges of the solarized areas, where control efficacy is usually lower relative to the center, a phenomenon termed "border effect" (Grinstein *et al.*, 1995). Solarization in combination with plant residues containing glucosinolates, such as broccoli and cabbage, generally improves nematode control efficacy of solarization (Gamliel & Stapleton, 1993; Stapleton & Duncan, 1998; Ploeg & Stapleton, 2001).

2. Hot water, steaming, flaming, and microwave irradiation

Hot water is another method used to disinfest soils by high temperatures. For example, drenching soil with 90°C water in the autumn or early spring reduced the foliar nematodes *Aphelenchoides fragariae* on hosta (*Hosta* spp.) in nurseries and landscapes (Jagdale & Grewal, 2004). In the late 1970s, a prototype for a hot-water application system was developed in Japan. The system applied hot water (70 to 95°C) onto the soil surface and raised the soil temperature to a level that was lethal to plant pathogens, pests and weed seeds (Takehara, 2004). Although relatively good control efficacy was obtained with this method, it may be only suitable for small areas due its water and heating energy requirements. Hot water is also used to control nematodes in planting materials, such as seeds, tubers, bulbs, rhizomes and corms (Bridge, 1996). Careful temperature management and exposure time is important to kill nematodes without killing the planting materials (bulbs, corms, seeds and tubers). Generally, planting materials are immersed in hot water in a range of 45 to 55°C for several minutes, but conditions vary depending on the planting material and nematode species.

Steaming, flaming and microwaving are other control methods based on soil-temperature elevation. Several methods are used for soil steaming, but the most effective one for soilborne diseases is the negative pressure system, in which steam is introduced between the soil surface and the steaming sheath and forced to enter the soil by negative pressure. The negative pressure is generated by a fan that sucks the air out of the soil through perforated polypropylene pipes installed at ca. 60 cm depth in the soil (Ruina, 1983). Using this method, the soil temperature is increased to 70 or 80°C, even at the depth of the installed tubes, effectively controlling nematodes and other soilborne pathogens and weeds. However, steaming is an expensive process due to its high energy costs, and it is not always considered "environmental friendly" because of CO_2 emission during steam generation by burning fuels. Pathogen reinfestation is usually more rapid in steamed soil than in solarized soil, probably due to the former's disruption of the biological balance of soil microflora. Generally, the maintenance of low nematode populations in steamed soils does not last for more than 2 or 3 years. Steaming every year is popular for high-value crops.

Flaming is a non-chemical control method used for weeds and pests. In Israel, an attachment which emits a flame jet has been developed for tractors (Margaliot Trade and Fumigation Ltd.), specifically for soil disinfestation. This flame jet is combined with a rotary cultivator, and the cultivated soil is exposed to the flame and high-temperature air. The soil temperature needs to be elevated to kill pathogens and weeds. This method is reported to be highly effective for weed control, but its efficacy for nematode control has not been fully proven. The practice of *slash and burn* combined with crop rotation and fallow, which is performed in some regions, may belong to this control method in certain respects. The practice has been reported to be effective in nematode control (Wilson & Caveness, 1980).

Microwaving is another technique used to kill nematodes by elevated temperatures. Pine wood nematodes were controlled in wood by using commercial 2.45-GHz microwave equipment (Fleming *et al.*, 2005). The irradiation resulted in 100% mortality of pine wood nematodes at temperatures measured at above 62°C. Although the experiments were done with wood, application to growth media and soils has been reported. Soilborne fungi and nematodes were effectively controlled using 2,450 ± 20 MHz waves (van Wambeke *et al.*, 1983), or similar microwaves, although the efficacy depended on the duration of irradiation (van Assche & Uyttebroeck, 1983). For field application of microwaves, high-energy input may be a limiting factor, in addition to soil conditions. Combinations with soil solarization are suggested to elevate the initial soil temperature, thereby saving on the energy required to generate microwaves and improving control efficacy (Mavrogianopoulos *et al.*, 2000). These methods, especially steaming and flaming, appear to be more destructive to the soil structure/micro- and macroflora than soil solarization because of the higher temperatures reached. These methods are not always permitted by the authorities in organic farming systems.

3. Flooding

Flooding in the field induces anaerobic conditions in the soil and kills pathogens, including nematodes (Stover, 1979). In addition, as mentioned in *Reductive soil disinfestation*, several organic acids which are toxic to nematodes are generated from organic materials in the soil under anaerobic conditions. This technique can be applied where water is plentiful, and the soil can sustain water for several weeks or months. Care should be taken that drainage from the nematode-infested field not spread nematodes to other fields. The flooding period may vary depending on the nematode species and their stage of development. Eggs are thought to be more tolerant to anaerobic conditions.

"Chemical" control with natural nematicides

1. Plant metabolites

Plants are a potential source of new chemicals for pesticide development. Nematicidal phytochemicals are generally safe for the environment and humans. Many nematicidal phytochemicals from a great variety of chemical structures, such as alkaloids, phenols, sesquiterpenes, diterpenes, polyacetylenes and thienyl derivatives, have been isolated from numerous plant families (Gommers & Bakker, 1988; Chitwood, 2002). Most of these nematicidal phytochemicals have been isolated from the family Asteraceae (Gommers & Bakker, 1988). Although many nematicidal compounds have been isolated from plants, their use for nematode control in agriculture as plant extracts or synthetic compounds based on natural nematicides is very limited, in contrast to insect control where natural nicotine and pyrethrins, and synthetic neonicotinoids and pyrethroids are used. The most practical method for applying nematicidal phytochemicals may be as green manure, as mentioned earlier, provided the plants can be grown in suitable seasons, they are not nematode hosts, and nematicidal compounds are released and active in the soil for an appropriate period of time. Leaf powder and organic solvent extracts of *Inula viscose* have been found to control *M. javanica* and other plant-parasitic nematodes (Oka *et al.*, 2001). Emulsifiable concentrate formulations of the pastes obtained from organic solvent extracts of the plant also killed *M. javanica* juveniles in soil at a concentration of 0.01% (w/w) or higher (Oka *et al.*, 2006a). Although the plant extract has potential as a nematicide, commercial products have not been further developed for several reasons. *Bidens pilosa*, another Asteraceae, has been confirmed to have nematicidal compounds, and its extracts have shown high nematicidal activity against *M. incognita* (Taba *et al.*, 2008).

Plant essential oils, which are volatile compounds extracted mainly from spices and aromatic plants, are commonly used in the medicinal, cosmetic and food industries. In addition to flavors and therapeutic activities, essential oils are well known to possess antimicrobial and insecticidal activities (Shaaya *et al.*, 1991; Isman, 1999). Some insecticides and insect repellents based on essential oil components have been commercially developed (Isman, 1999). Essential oils of some plants and their components have been tested for their nematicidal activity *in vitro* and in soil (Sangwan *et al.*, 1990; Walker & Melin, 1996). Essential oils of applemint (*Mentha rotundifolia*), caraway (*Carum carvi*), fennel (*Foeniculum vulgare*), oregano (*Origanum vulgare*), Syrian oregano (*Origanum syriacum*) and wild thyme (*Coridothymus capitatus*) were found to have nematicidal activity against *M. javanica* at concentrations of ≥125 μl/l *in vitro* and ≥100 mg/kg in the soil (Oka *et al.*, 2000b). The main components of these essential oils, such as carvacrol, thymol, *t*-anethole and (+) and (-)-carvone, were also found to be nematicidal. Essential oils of garlic and *Cinnamomum verum* were found to be effective against the pine wood nematode, *Bursaphelenchus xylophilus* (Park *et al.*, 2005). Although essential oils are effective in nematode control under experimental conditions, their cost is prohibitive for field applications. Since chemically synthesized essential oil components cannot be used in organic farming systems, plants containing nematicidal essential oils may be used as green manures and for biofumigation.

There are some commercial natural nematicides based on plant extracts (Table 4). Extracts of *Quillaja saponaria*, which is known to contain saponins and polyphenols, have been used to control nematodes in Chile (San Martin & Magunacelaya, 2005). Field trials showed that the use of 30 l/ha of the formulation QL 1000® (Natural Response, Chile) controlled nematodes such as *Meloidogyne* and *Xiphinema* spp. in vineyards and table grapes as effectively as synthetic nematicides. Several commercial natural nematicides have been developed from plant extracts, which may also improve root growth. Sincocin® (Agricultural Sciences, Texas) is a liquid concentrate based on plant extracts and fatty acids that reduces the feeding vigor and consequent damage of plant-parasitic nematode species, such as of the citrus nematode (*T. semipenetrans*) on orange, the reniform nematode (*R. reniformis*) on sunflower, and the sugar beet cyst nematode (*H. schachtii*) on sugar beet, and stimulates certain predatory nematode species that are important in the biological control of plant-parasitic nematode populations. The product significantly reduced the citrus nematode population and enhanced the growth of orange trees (Osman & Salem, 1995). Mustard and chili pepper extracts are also used for nematode and soilborne disease control. Dazitol® (Champon Millennium Chemicals, Virginia), which is a natural soil fumigant containing a blend of extracts including chili and mustard, controls plant-parasitic nematodes, including *Xiphinema americanum*, *Meloidogyne* spp., *Pratylenchus*, *Trichodorus*, *Tylenchus*, *Ciconemoides*, and *Paratylenchus* spp. Natural nematicides based on sesame oil and sesame plants are also available. Neotrol Liquid® (Barmac Industries, Queensland, Australia), which is derived from extracts of specific cultivars of hybrid sesame plants, is used for nematode control. Dragonfire® (NaEx Corp.-Poulenger, Texas) is another product based on the oil of a specific cultivar of wild sesame. As already mentioned, several commercial formulations based on neem extracts have been developed in some countries, mainly for insect control; however, most commercial neem products have nematicidal activity as well (Reshmi & Vijayalakshmi, 1998). As compared with synthetic chemical nematicides, these natural nematicides generally have lower control efficacies (Crow, 2005), and if high nematode populations are monitored in a field, other control means should be used in combination with them to obtain satisfactory results. An advantage of most of these natural nematicides is that they can be applied during the growing season in the field without leaving chemical residues in the yield, provided the products are not phototoxic.

2. Microbial metabolites

Microorganisms are another source of nematicidal compounds (Table 4). Several microorganism metabolites have been discovered to have nematicidal activity. Avermectins, a class of macrocyclic lactones produced by *Streptomyces avermitilis*, have long been known to be toxic to insects, mites and nematodes, and are used as insecticides and acaricides, as well as anthelminthics for domestic animals (Jansson & Dybas, 1998). Abamectin (avermectin B_{1a} and B_{1b}) has been used to control several plant-parasitic nematodes (Sasser *et al.*, 1982; Garabedian & Van Gundy, 1983), but these compounds have not been successfully developed as nematicides for soil application due to their insolubility in water and their high affinity to soil particles and organic matter. Nevertheless, seed-dressing agents based on abamectin (Avicta®, Syngenta, Switzerland) have been recently developed for nematode control (although not for organic farms), specifically to protect cotton and maize seedlings from infection with root-knot and other nematodes in the early growth stage. Although avermectins are natural compounds, they are not permitted for use in organic farms due to their antibiotic feature. DiTera® (Valent Biosciences, Illinois), which is based on metabolites and killed mycelia of the fungus *Myrothecium verrucaria*, has been developed as a natural nematicide. The product controls several economically important nematode species, such as root-knot, cyst, and lesion nematodes in the soil (Warrior *et al.*, 1999). The product is composed primarily of proteins, sugars, and lipids and is stable under heat and environmental rigors. Use of this product is permitted in organic farms in a few countries.

Entomopathogenic nematodes, which are used as biological control agents of insects, carry symbiotic bacteria in their bodies; for instance, *Xenorhabdus* spp. are carried by *Steinernema* spp., and *Photorhabdus* spp. by *Heterorhabditis* spp. Metabolites such as 3,5-dihydroxy-4-isopropylstilbene and indole, isolated from the bacteria *P. luminescens*, are toxic to *M. incognita* (Hu *et al.*, 1999). Similarly, *Xenorhabdus* spp. and their metabolites suppressed hatch in *Meloidogyne* spp., and in some instances caused mortality of juveniles (Hu *et al.*, 1995; Samaliev *et al.*, 2000). Bacterial metabolites have never been used for field applications. However, entomopathogenic nematodes have been used to control plant-pathogenic nematodes (see *Biological control* section in this chapter).

Table 4. Some commercial nematicides based on natural compounds

Product®	Main constituent	Main target nematode	Company
BioAct	Fungal spores of *Paecilomyces lilacinus*	*Meloidogyne, Heterodera, Globodera, Pratylenchus, Radopholus similis*	Prophyta, Germany
BioNem, BioSafe	Bacterial spores of *Bacillus firmus*	*Meloidogyne, Heterodera, Helicotylenchus, Hoplolaimus, Trichodorus, Tylenchulus semipenetrans, Xiphinema index, Ditylenchus dipsaci*	Agrogreen, Israel
Ne-Plex	Bacterial spores of *Bacillus chitinosporus*	Plant-parasitic nematodes	Terra Biotics, Inc., New Mexico, USA
DiTera	Metabolites and dead mycelium of the fungus *Myrothecium verrucaria*	Root-knot, cyst, lesion, stubby root, sting and several other nematodes	Valent BioSciences, Illinois, USA
Dazitol	Mustard and chili pepper extracts	*Xiphinema americanum, Meloidogyne* spp., *Pratylenchus* spp., *Trichodorus, Tylenchus, Pratylenchus, Criconemoides, Paratylenchus*, etc.	Champon Millennium Chemicals, Virginia, USA
Sincocin	Several plant extracts	Plant-parasitic nematodes	Agricultural Sciences, Texas, USA
QL Agri	Plant extract of *Quillaja saponaria*	Plant-parasitic nematodes	Desert King International, California, USA
Neotrol Liquid	Extracts of sesame plants	Lance, string, root-knot and cyst nematodes	Barmac Industries, Queensland, Australia
Dragonfire-CPP	Oil of wild sesame seeds	Plant-parasitic nematodes in turf and lawns	NaEx Corp-Poulenger, Texas, USA
Nemagard	Ground sesame stalks	Various plant nematodes such as root-knot, root-lesion, citrus, lance, stubby-root and dagger nematodes.	Natural Organic Products, Florida, USA
CA-1	Oriental mustard seeds containing allyl isothiocyanate	Nematodes in turf and ornamentals	Nematrol, Inc., Ontario, Canada
Neemate 10 G	Azadirachtin and other Neem-based triterpenoids	Plant-parasitic nematodes	Nico Orgo USA, California, USA
Ozoneem Cake	Neem cake	Plant-parasitic nematodes	Ozone Biotech, Haryana, India

Several microbial metabolites have been isolated from the culture media of bacteria and fungi in the laboratory; some are general compounds with nematicidal activity, such as ammonia, nitrous acid and acetic acid, while others have not been identified, or include only a few identified compounds. Commercialization of microbial metabolites as nematicides will require more R&D by companies and public laboratories.

Biological control

Biological control is generally less consistent, less effective and slower-acting than chemical control. Nematode control by biological agents or antagonistic organisms can be performed via two main approaches: 1) introduction of (hyper-) parasitic or antagonistic organisms to nematode-infested soils, 2) augmentation of indigenous antagonistic organisms by manipulating the soil environment, such as by incorporating organic amendments into the soil (Stirling, 1991; Sikora, 1992). Many parasites of nematodes and organisms that are antagonistic to nematodes have been found in the soil, mainly fungi, bacteria and predatory nematodes (Fig. 7) (Jatala, 1986; Stirling, 1991). Mass production of some of these natural enemies/antagonists is relatively easy because of their facultative parasitic features, but their introduction into fields does not always result in satisfactory control; on the other hand, other promising natural enemies can be propagated only on their prey, i.e. live nematodes. Mass production and commercialization of such obligate parasites or predators is therefore difficult. Moreover, nematode control efficacy of biological agents is greatly influenced by soil environments, and some control agents are highly specific to particular nematode species. Nevetheless, there are a few commercial nematode-control products based on natural enemies or antagonists (Table 4).

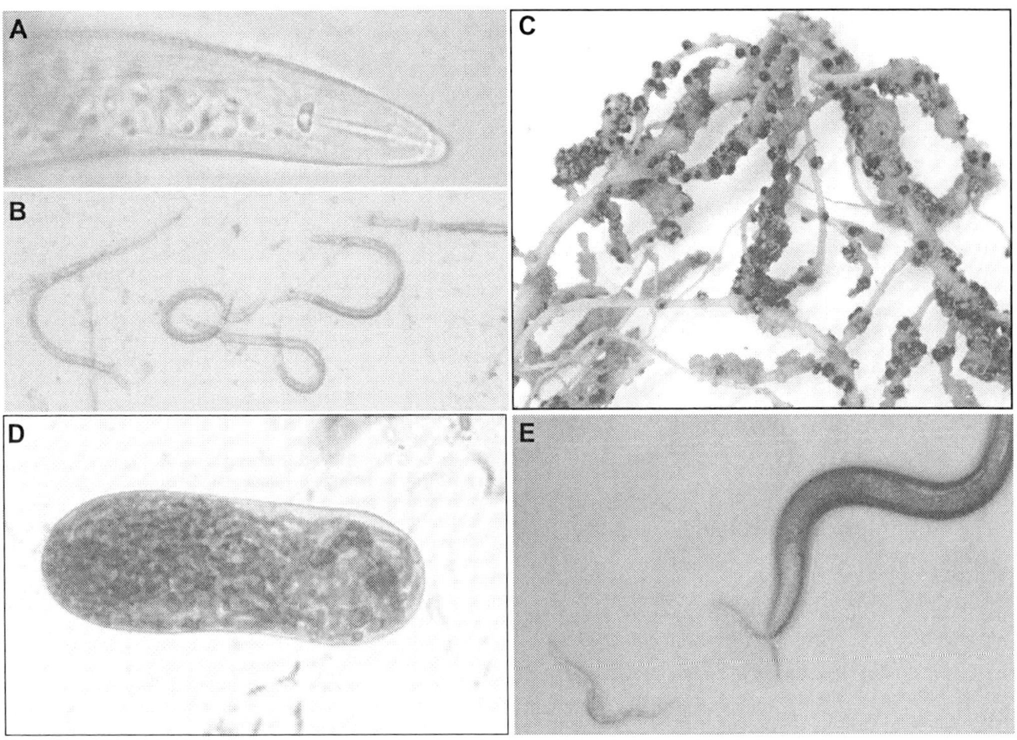

Fig. 7. Natural enemies of nematodes. A: Spores of *Pasteuria penetrans* (arrows) inside the body of *Helicotylenchus* sp. B: *Meloidogyne* juveniles trapped by *Arthrobotrys* sp. mycelium. C: *M. incognita* egg masses infected by the *Scytalidium*-like fungus on pepper roots. Infected egg masses are black whereas uninfected ones are white to pale-brown. D: A *M. incognita* egg infected with the *Scytalidium*-like fungus. The egg content has been replaced by fungal mycelia. E: The predatory nematode *Koerneria sudhausi* feeding on a *Meloidogyne* juvenile.

1. Bacterial antagonists

The most studied and promising bacterium for nematode control is *Pasteuria penetrans*, which is a very specific obligate parasite on nematodes (Fig. 7A) (Chen & Dickson, 1998, 2004). *Pasteuria* species are gram-positive, endospore-forming bacteria, which are genetically close to *Bacillus* and *Clostridium*. The bacteria preferentially grow inside the body of developing nematodes, and inhibit nematode reproduction. A few species of *Pasteuria* have been identified from nematodes: *P. nishizawae* from *Heterodera* and *Globodera* spp., *P. thorney* from *Pratylenchus* spp., and *P. penetrans* from *Meloidogyne* spp. Other proposed species have been isolated from other nematode species. These bacterial species have been isolated from nematode-suppressive soils in several places in the world (Stirling & White, 1982; Minton & Sayer, 1989; Dickson *et al*., 1994). The bacterium has been recognized as a potential biocontrol agent for root-knot nematodes for more than two decades. Application of the bacterial spores, which had been cultivated on root-knot nematodes on tomato plants, effectively reduced root galls caused by *M. javanica* (Stirling, 1984). In fact many experiments have been performed with these bacteria, showing their potential for nematode control in many plants (Chen & Dickson, 1998). Experiments using *P. penetrans* against *M. arenaria* on peanut in Florida indicated conditions for successful control of the nematodes over several years: high concentrations (ca. 100,000 per gram soil) of the spores need to be applied. If lower concentrations are used, susceptible crops must be planted for consecutive years to increase the bacteria in the nematodes. The nematode population will be maintained at a non-damaging level when the bacterial spores reach concentrations above 133,000 spores per gram soil (Chen & Dickson, 2004). The obligate nature of the bacteria's parasitism has inhibited progress with large-scale culture and product deployment. However, a commercial product based on *P. penetrans* spores (Pasteuria®, Nematech, Japan) has been released to the market for the control of major species of root-knot nematodes in Japan. The product contains a few strains of the bacteria isolated from these major root-knot nematode species. Because the spores are produced on developing root-knot nematodes in plant roots, the price of the product is very high for field applications (ca. $50,000/ha). Susceptible crops must be grown in the treated field for a few years to increase and establish the bacterial population in the soil. A breakthrough in the mass production of *P. penetrans* is *in-vitro* culture (Hewlett *et al.,* 2002). A venture company in the US announced that they had succeeded to produce an *in-vitro* culture of the bacterium, and that in a few years, they plan to release a commercial product.

Another bacterial agent for nematode control which has been released commercially is a formulation of the bacterial spores of *Bacillus firmus* (BioNem®, Agro-green, Israel). The product has been reported to be effective against root-knot and cyst nematodes, the citrus nematode *T. semipenetrans*, the dagger nematode *Xiphinema index* and the stem and bulb nematode *Ditylenchus dipsaci* (Keren-Zur *et al*., 2000). A formulation applied at 200 and 400 kg/ha was effective at reducing the number of galls and increasing tomato shoot height and weight relative to untreated controls, 45 days after treatment in small plots (Terefe *et al*., 2009). The product significantly reduced nematode populations and root infestation in experiments on vegetable crops, including tomato, cucumber, pepper, garlic and some herbs (Giannakou & Prophetou-Athanasiadou, 2004; Giannakou *et al*., 2007). The producer reports that the bacteria degrade eggs of root-knot nematodes. Soil treatments with the bacteria occasionally improves plant growth, probably due to the fact that the species is known as a phytohormone-producing and phosphate-solubilizing bacterium, and the additives in the formulations serve as fertilizer. A formulation that is applicable via irrigation tubes is also available, in addition to a granulate formulation for soil incorporation.

In addition to the direct parasitism of bacteria on nematodes, some bacteria are known to reduce nematode infection and/or increase plant resistance to nematodes. Some rhizobacteria, termed "plant growth-promoting rhizobacteria" (PGPR), have been reported to increase plant growth. Some reports have indicated that use of PGPR in seedling media improves plant growth and occasionally reduces damage caused by root-knot nematodes, as well as other soilborne diseases. These bacteria are mainly rhizobacteria, such as *Bacillus* and *Pseudomonas* spp. (Kluepfel *et al.*, 2002). For example, root galls caused by *M. incognita* on pepper and mask melon were reduced by some strains of PGPR (Kokalis-Burelle *et al.*, 2002, 2003), and damage on coffee plants by *M. incognita* was reduced by *Bacillus* spp. endophytes, such as *B. pumilus* and *B. mycoides* (Mekete *et al.*, 2009). A study indicated that rhizobacteria from antagonistic plant roots, such as velvet bean, castor bean and rye, were mainly gram-negative genera, whereas those isolated from soybean roots were mainly *Bacillus* spp. (Kloepper *et al.*, 1992), and a higher percentage of rhizobacteria isolated from the antagonistic plants reduced nematode damage on soybean by *H. glycines* and *M. incognita* than those isolated from the soybean rhizosphere. The mechanism by which rhizobacteria and bacterial endophytes reduce nematode infection is not clearly understood, but they may produce nematicidal compounds (antibiotics), inhibit nematode orientation towards the host roots by modifying root exudates, and elicit the plant defense system known as systemic acquired resistance (Reitz *et al.*, 2000). Although several experiments have shown their potential as nematode control agents, no commercial products based on these bacteria have been developed.

2. Fungal antagonists

Many fungi have been known to infect vermiform nematodes by forming traps (Fig. 7B), including adhesive nets, rings, knobs, and conidia and constricting rings, including *Arthrobotrys* spp., *Dactylaria* spp., *Dactylella* spp., *Hirsutella* spp. and *Monacrosporium* spp. (Gray, 1988; Chen & Dickson, 2004). These fungi are facultative parasitic fungi that can be easily grown and propagated on artificial media. Several attempts have been made to develop nematode control products based on these parasitic fungi. However, satisfactory control was not always obtained, probably due to low rates of trap formation, which may strongly depend on nematode density and the fungi's establishment in the soil (Jaffee, 2004; Liu & Chen, 2005).

Several fungi are known to infect cysts and nematode eggs (Morgan-Jones & Rodríguez-Kábana, 1988; Chen & Dickson, 2004). The fungus *Pochonia chlamydosporia* (=*Verticillium chlamydosporium*) has been studied for its potential as a biocontrol agent because this facultative parasitic fungus can be grown casily, and infects females and eggs of cyst, root-knot and other nematode species (de Leij & Kerry, 1991; Kerry & Jaffee, 1997). Field experiments with granular formulations containing *V. chlamydosporium* or *Arthrobotrys dactyloides* in Australia showed that formulations of *A. dactyloides* at 220-440 kg/ha substantially reduce the number of root-knot nematodes present in roots 4 to 8 weeks after planting, whereas those of *V. chlamydosporium* do not increase egg parasitism or reduce galling or nematode numbers at harvest (Stirling & Smith, 1998). One of the difficulties in developing commercial products using latter fungus is its genetic diversity and characteristics among individual isolates that depend on the region and infecting nematode species. Thus careful selection of isolates is required for target nematodes (van Damme *et al.*, 2005). A selected isolate of *P. chlamydosporia* var. *catenulata* was applied at densities similar to those that occur naturally in nematode-suppressive soils (Atkins *et al.*, 2003). The fungus significantly reduced infestations of *M. incognita* on tomato crops when used in combination with crop rotation. The survival of the fungus in soil was checked by a molecular method using specific primers for the fungus. However, a commercial product based on spores of *P. chlamydosporia* has only been released in Cuba for root-knot nematodes.

Another facultative parasitic fungus is *Paecilomyces lilacinus,* which can also infect females and eggs of cyst, root-knot and other nematodes (Jatala, 1986). Several strains of the fungus were tested against root-knot nematodes (Cabanillas & Barker, 1989), and one strain from the Philippines (strain PL 251) was developed for commercial use. Although several strains of the fungus have been isolated from human inflectional diseases, the isolate used as the nematode control agent is safe to humans because it does not produce mycotoxins, and cannot grow at a temperature of 36°C. Acute toxicology and pathogenicity studies indicate that *P. lilacinus* strain PL 251 has negligible to no mammalian toxicity. The commercial product based on the fungus (BioAct®, Prophyta, Germany) is said to be effective at infecting and destroying nematode eggs, as well as vermiform root-knot, cyst, burrowing (*Radopholus* spp.) and lesion nematodes (Kiewnick & Sikora, 2006). A few products based on the same fungus have been developed for nematode control by other companies and are commercially available. Combinations of egg and vermiform parasites may be more effective in nematode control because each fungal parasite attacks different stages of the target nematode (Khan *et al.*, 2006). Other fungi, including identified and unidentified fungi such as *Fusarium oxosporum*, ARF18 and a *Scytalidium*-like fungus (Fig. 7C,D), have been isolated from females and eggs of root-knot and cyst nematodes; however, no products based on these fungi have been developed for release in the marketplace (Kim & Riggs, 1991; Chen *et al.*, 1996; Oka *et al.,* 1997). Their commercialization requires further studies.

Similar to bacterial antagonists, some fungi, mainly fungal endophytes, increase the plant's resistance to nematodes (Sikora *et al.,* 2008). The presence of the fungal endophyte *Neotyphodium coenophialum* in tall fescue confers resistance to some plant-parasitic nematodes (Timper *et al.*, 2005). Non-pathogenic *Fusarium* was reported to play a role in resistance to root-knot nematodes in tomato (Hallmann & Sikora, 1994; Dababat & Sikora, 2007). Possible mechanisms involved in the increased resistance conferred by endophytes are: 1) inhibiting nematode infection by physically occupying the infection or feeding sites; 2) triggering plant defense responses against pathogens; 3) producing metabolites that are toxic to nematodes, and 4) interfering with the nematodes' host-searching behavior by modifying plant root exudates. Metabolites of a non-pathogenic *F. oxysporum* strain isolated from tomato roots were toxic to *M. incognita* and some soilborne fungi (Hallmann & Sikora, 1996). The fungus *Trichoderma* has been developed as a biocontrol agent for the control of soilborne fungi, such as *Pythium* spp., *Rhizoctonia solani*, *Fusarium* spp. and *Sclerotium rolfsii,* and as a growth enhancer of crop seedlings (Chet, 1990). Several *Trichoderma* spp., including *T. harzianum*, *T. koningii* and *T. lignorum*, are used to control plant-parasitic nematodes (Spiegel & Chet, 1998; Windham *et al.*, 1989). In greenhouse experiments, soil treatment with *T. harzianum* peat-bran preparations reduced root galling and increased top fresh weight of tomatoes grown in *M. javanica*-infested soil. The fungus showed the ability to infect *M. javanica* eggs and juveniles in the laboratory (Sharon *et al.*, 2001). These nematode parasites and inducers of nematode resistance may be applied in the nursery as a component of the seedling growth media to protect plants from nematode infection and promote plant growth in the field.

3. Predatory and entomopathogenic nematodes

The idea of using predatory nematodes to control plant-parasitic nematodes was suggested almost 90 years ago (Cobb, 1920). Predatory nematodes from three orders, Dorylaimida, Mononchida and Diplogasterida, are well known, and several attempts to use these representatives to control plant-parasitic nematodes have been made (Cohn & Mordechai, 1974; Khan & Kim, 2005). In contrast to entomopathogenic nematodes that have been developed into commercial biocontrol agents for insect pests, no commercial products exploiting predatory nematodes against plant-parasitic nematodes have yet been developed. There are several limitations to using predatory nematodes as biological control agents: difficulties in mass propagation of most predatory nematodes because of their low fecundity and long life cycle, and their need for live prey (nematodes) as a food source (Bilgrami & Brey, 2005). Moreover, the lack of a resistant stage for surviving under unfavorable environmental conditions and a high rate of cannibalism also make their propagation and maintenance difficult. A series of studies on the diplogasterids *Mononchoides longicaudatus* and *M. fortidens* suggested that diplogasterid predators are good candidates as biocontrol agents for several reasons: 1) they can be cultured on bacteria; 2) their life cycle is short; 3) they have a high rate of predation and a broad host range; 4) they have a juvenile stage (as dauer larvae) that is tolerant to adverse conditions, and 5) they show a low rate of cannibalism (Bilgrami & Brey, 2005). In a few studies, the efficacy of predation on plant-parasitic nematodes by diplogasterid predators was studied in a soil environment. *M. fortidens* and *M. longicaudatus*, which had been cultured on prey nematodes, yielded a reduced root-galling index on tomato plants grown in root-knot-nematode-infested soil in pots (Fauzia *et al.*, 1998; Khan & Kim, 2005). Recently, two diplogasterids, *Koerneria sudhausi* (Fig. 7E) (Bar-Eyal *et al.*, 2008) and *Mononchoides gaugleri* (Bilgrami et al., 2008), were propagated on bacteria rather than prey nematodes, and evaluated in pots and in the field, respectively. The oviposition rate of a female *K. sudhausi* feeding on bacteria in a Petri dish was about 6.7 eggs per day at 25°C without mating, and its life cycle was 8 days at the same temperature. The nematodes, introduced into *M. javanica*-inoculated soil before tomato transplanting, reduced the root-galling index in pot experiments. The field release of *M. gaugleri* resulted in significant control of plant-parasitic nematodes in a turf grass field (Bilgrami *et al.*, 2008). These results suggested that the development of predatory nematodes as biocontrol agents against plant-parasitic nematodes is not impossible.

The entomopathogenic nematodes *Steinernema* spp. and *Heterorhabditis* spp. have been produced commercially for insect pest control. Antagonistic effects of the entomopathogenic nematode-bacterium complexes on a range of plant-parasitic nematodes have been demonstrated (Grewal *et al.*, 1997; Fallon *et al.*, 2004) but research has been confined mainly to laboratory studies. Because these entomopathogenic nematodes are not predatory to nematodes, their mode of action in the control of plant-parasitic nematodes is not understood. However, some experiments suggest that the entomopathogenic nematode *Steinernema carpocapsae* and the fungivorous nematode *Aphelenchus avenae* are attracted to the root tips, and their application reduced root galls on tomato caused by *M. incognita* (Ishibashi & Choi, 1991). The root tip is the infection site of *Meloidogyne* spp. as well as some other nematode species. Metabolites of the symbiont bacteria of the entomopathogenic nematodes are reported to be toxic to some plant-parasitic nematodes (Hu *et al.*, 1999). However, another study showed ineffective control of *M. javanica* on tomato using *Steinernema feltiae* in pot experiments (Fallon *et al.*, 2002). Although the entomopathogenic nematodes are commercially available as biocontrol agents for insect pests, more research is needed to obtain satisfactory control efficacy by these entomopathogenic nematodes.

4. Increase in antagonists by organic amendments

Linford (1937) and Linford *et al.* (1938) observed that incorporation of organic amendments enhanced nematode-antagonistic flora and fauna, and reduced populations of plant-parasitic nematodes in the soil. Several organic amendments, some applicable in fields, others used only for experimental purposes, were tested for how they affect nematodes and nematodes' antagonists (Stirling, 1991; D'Addabbo, 1995). Antagonists of nematodes include trap-forming fungal parasites, nematode egg parasites, parasitic bacteria, microorganisms producing hydrolytic enzymes that are deleterious to nematodes, predatory nematodes, and other invertebrates (Stirling, 1991). Incorporation of sunnhemp (*Crotalaria juncea*) to soil increased nematode-trapping fungi, parasitic fungi on *Rotylenchulus reniformis* eggs, vermiform-stage parasites, and bacterivorous nematodes more efficiently than amendments with *Brassica napus* or *Tagetes erecta* (Wang *et al.*, 2001). Among the many kinds of organic amendments, chitinous materials have attracted special attention for their control of soilborne diseases and nematodes because chitin is present in nematode eggs as well as in the cell walls of most plant-pathogenic fungi (Mian *et al.*, 1982; Spiegel *et al.*, 1987; Gotlieb *et al.*, 2003; De Jin *et al.*, 2005). Decomposition of chitin by soil microorganisms causes the release of volatiles, mainly ammonia, that are toxic to pathogens (Spiegel *et al.*, 1987). This control effect is transient and occurs only in the early period after amendment incorporation. At the same time, the number of microorganisms with chitinolytic activity (i.e., those producing chitinases) and the amount of chitinase activity in the soil increase (Spiegel *et al.*, 1987; De Jin *et al.*, 2005). The prolonged effect of nematode control is believed to be due to the increase in antagonistic microorganisms (Spiegel *et al.*, 1987). Some microorganisms isolated from chitin-amended soil were tested for their biocontrol efficacy against nematodes (Spiegel *et al.*, 1991; Oka *et al.*, 1993; Cronin *et al.*, 1997; Tian *et al.*, 2000). Among them, the newly identified bacterium *Telluria chitinolytica* (=*Pseudomonas chitinolytica*) showed potential for root-knot nematode control (Spiegel *et al.*, 1991). A unique amendment for nematode control that has been tested in the laboratory is collagen, which is the main protein of nematode cuticles (Galper *et al.*, 1990). Although the ammonia concentration in collagen-amended soil was lower than the concentration that is lethal to *M. javanica*, the amendment reduced nematode damage on tomato. A collagenolytic fungus, *Cunninghamella elegans*, was isolated from the soil, and nematicidal activities of enzymes produced by the fungus were demonstrated (Galper *et al.*, 1991). Application of one or a few microorganisms isolated from amended soil may not be enough for satisfactory control of nematodes. To increase the microorganisms' nematode control activity, a combination of isolated microorganisms and amendments that serve as substrates for them and may help in their establishment in the soil is preferred (Galper *et al.*, 1991; Oka *et al.*, 1993; Chen *et al.*, 1999; Tian *et al.*, 2000).

5. Naturally occurring biological control

Nematode damage has been found in several places in the world to remain below the economic threshold after a history of nematode damage in the same field. Such nematode-suppressive soils are usually obtained after growing nematode-susceptible plants for several years (Stirling, 1991). Natural enemies or antagonists have been occasionally isolated from such nematode-suppressive fields, and their potential for biological control has been evaluated. For example, damage of cereals by the cyst nematode *H. avenae* in northern Europe is not common under monoculture of susceptible cereals, whereas the nematode is a severe problem for cereals grown in Mediterranean climates, such as in Australia and Mediterranean countries, even under monoculture conditions. Very high percentages of nematode eggs and females were infected with parasitic fungi, such as *Nematophthora gynophila* and *P. chlamydosporium*, in nematode-suppressive cereal fields in the UK (Kerry, 1975). Natural suppression of the sugar beet cyst nematode *H. schachtii* was observed in California, where nematode-parasitic fungi have been isolated. One such fungus is *Dactylella oviparasitica*, whose nematode control efficacy was confirmed in several field experiments (Olatinwo *et al.*, 2006). The fungus was also thought to be the main factor determining the low root-knot nematode population in an old peach orchard in California (Stirling *et al.*, 1979). In a soybean field in Arkansas, the population density of the soybean cyst nematode *H. glycines* was decreased naturally by the sterile fungus ARF18 (Kim & Riggs, 1991). The parasitic bacterium *P. penetrans* has been suggested as the main factor in root-knot nematode-suppressive soils in several locations in the US, where nematode damage declined after monoculture (Minton & Sayre, 1989; Weibelzahl-Fulton *et al.*, 1996). Introduction of suppressive soil to other soils makes the latter soils suppressive in experiments, but the practical application of this finding for fields is very limited. Introduction of the natural enemies existing in nematode-suppressive soils into other fields only barely succeeds in nematode control, because establishment of the natural enemy is greatly affected by soil conditions and other factors.

One important question is: if nematode-susceptible crops are grown for several years without controlling nematodes, will the field become nematode-suppressive, possibly due to the build-up of natural enemies and antagonists of the nematode in the soil? General opinion is that the population of plant-parasitic nematodes in the field soil will not always be kept under the threshold of nematode economic damage, even after cropping nematode-susceptible crops, where the preferential establishment of a nematode-suppressive soil may be rare. The balance of nematode populations in soil microflora is not determined simply by their natural enemies, but also by several other factors, including, climate, soil characteristics, crops, cultural practices, etc.

Integrated management

Applying a single control measure for nematode control in organic farming systems, and even in conventional agriculture, does not generally produce satisfactory results. Application of combinations of the above-described control methods to the entire agrosystem is essential to reducing nematode damage, wherever possible in terms of cost, feasibility and time (Oka *et al.*, 2000a; Atkins *et al.*, 2003). Nematode control is much more difficult during the crop-growing period than before planting or seeding. Maximum effort should be made to reduce nematode populations in the soil before the growing season. For example, when the nematode infection level is high, and growth of a high-value crop is planned, a combination of cover crop-biofumigation/solarization-resistant varieties or grafted plants may be effective. If there is not enough time to grow a cover crop, more intensive methods are required, such as steaming or combinations of ammonia-releasing amendments and solarization before planting. During the growing season, applicable nematode control methods are very limited, such as application of biocontrol agents and/or natural nematicides, and the control efficacy is not always satisfactory or cost-effective. Nematodes are only one group of soilborne pathogens, and control of other soilborne diseases and weeds also has to be considered. Fortunately, many of the control methods mentioned here are also effective for other soilborne diseases.

Conclusion

Soilborne diseases are very difficult to control due to the pathogens' biological niche. Nematodes are the most "problematic" soilborne pathogens from the standpoint of their diversity, mobility in the soil, lack of available knowledge and a scarcity of available nematicides. Nematodes are generally thought to be less incident in organic soils—the most important element of organic farming, where farmers make vast efforts to "build" such soils over many years. However, this assumption is not always correct. Intensive organic farming systems in greenhouses, where crop rotations with antagonistic cover crops are not feasible for economic reasons, are likely to have more nematode problems, and greater consequent economic losses.

Here, several control methods are described, but the most important and easiest control for nematodes in the field is preventing their introduction in the first place, or preventing their spread. Developments of quarantine systems between countries or states, and observation of nurseries by authorities result in reduced nematode spread via planting materials, but farmers should always be aware of the danger and careful with their materials. Increasing the farmers' awareness of nematodes via extension services may also reduce nematode spread not only in developing countries, but also in developed ones. Some control methods, such as soil solarization and organic amendments, can be used for all types of nematodes, while most of the methods herein require identification of the nematodes in the field, and an understanding of their biology (life cycle, host range, survival strategies, etc.) before application. Farmers need to perform sampling for identification and to consult experts or extension services. Eradication of nematodes at the field level is almost impossible; however, farmers can maintain their population densities below economic threshold by combining control methods.

Since restrictions on the use of nematicides and soil fumigants in agriculture have become more intense worldwide, nematode control methods that can also be used in organic farming are expected to undergo more intensive development. More "environmentally friendly" nematicides based on plant extracts and tissues, microbial metabolites and antagonistic microorganisms should be released in the near future. Seed companies have been putting more effort into investing the development of nematode-resistant varieties and rootstocks. These advances will help organic farmers manage nematodes, support sustainable agriculture and protect the environment, while supplying safer food to consumers.

REFERENCES

Aballay, E., Sepulveda, R., & Insunza, V. (2004). Evaluation of five nematode-antagonistic plants used as green manure to control *Xiphinema index* Thorne et Allen on *Vitis vinifera* L. *Nematropica 34*, 45-51.

Abid, M., & Maqbool, M. A. (1990). Effects of intercropping of *Tagetes erecta* on root-knot disease and growth of tomato. *International Nematology Network Newsletter 7*, 41-42.

Akhtar, M. (1998). Biological control of plant-parasitic nematodes by neem products in agricultural soil. *Applied Soil Ecology 7*, 219-223.

Alcaniz, E., Pinochet, J., Fernandez, C., Esmenjaud, D., & Felipe, A. (1996). Evaluation of rootstocks for root-lesion nematode resistance. *HortScience 31*, 1013-1016.

Alexander, S. A., & Waldenmaier, C. M. (2002). Suppression of *Pratylenchus penetrans* populations in potato and tomato using African marigold. *Journal of Nematology 34*, 130-134.

Ali, M., Matsuzoe, N., Okubo, H., & Fujieda, K. (1992). Resistance to non-tuberous *Solanum* to root-knot nematode. *Journal of Japanese Society for Horticultural Science 60*, 921-926.

Anwar, S. A., & McKenry, M. V. (2002). Developmental response of a resistance-breaking population of *Meloidogyne arenaria* on *Vitis* spp. *Journal of Nematology 34*, 28-33.

Araki, M., Inoue, T., & Sudo, M. (1991). [The effects of marketed forage crops on the population of root-knot nematode, *Meloidogyne incognita*, under field conditions.] *Proceedings of the Association for Plant Protection of Kyushu 37*, 130-132.

Atkins, S. D., Hidalgo-Diaz, L., Kalisz, H., Mauchline, T. H., Hirsh, P. R., & Kerry, B. R. (2003). Development of a new management strategy for the control of root-knot nematodes (*Meloidogyne* spp.) in organic vegetable production. *Pest Management Science, 59*, 183-189.

Atu, U. G. (1984). Effect of cover plants in fallow lands on root knot nematode population. *Beitrage zur Tropischen Landwirtschaft und Veterinarmedizin 22*, 275-280.

Aung, T., & Prot, J. C. (1990). Effect of crop rotation on *Pratylenchus zeae* and on yield of rice cultivar UPL Ri-5. *Revue de Nématologie 13*, 445-447.

Ayala, A. (1971). Rotation with pangola grass and sugarcane beneficial for pineapple production in Puerto Rico. *Nematropica 1*, 6-7.

Bachthaler, G., & Hien, L. (1976). [Results of long-term cereal rotation with varying proportions of maize.] *Bayerisches Landwirtschaftliches Jahrbuch 53*, 387-403.

Bar-Eyal, M., Sharon, E., & Spiegel, Y. (2006). Nematicidal activity of *Chrysanthemum coronarium* L. *European Journal of Plant Pathology 114*, 427-433.

Bar-Eyal, M., Sharon, E., Spiegel, Y., & Oka, Y. (2008). Laboratory studies on the biocontrol potential of the predatory nematode *Koerneria sudhausi* (Nematoda: Diplogastridae). *Nematology 10*, 633-637.

Bélair, G. (1992). Effects of cropping sequences on population densities of *Meloidogyne hapla* and carrot yield in organic soil. *Journal of Nematology 24*, 450-456.

Bélair, G., Dauphinais, N., Fournier, Y., & Dangi, O. P. (2004). Pearl millet for the management of *Pratylenchus penetrans* in flue-cured tobacco in Quebec. *Plant Disease 88*, 989-992.

Belcher, J. V., & Hussey, R. S. (1977). Influence *of Tagetes patula* and *Arachis hypogaea* on *Meloidogyne incognita. Plant Disease Reporter 61*, 525-528.
Bellostas, N., Sørensen, J. C., & Sørensen, H. (2007). Profiling glucosinolates in vegetative and reproductive tissues of four *Brassica* species of the U-triangle for their biofumigation potential. *Journal of the Science of Food and Agriculture 87*, 1586-1594.
Bengtsson, J., Ahnstrom, J., & Weibull, A.- C. (2005). The effect of organic agriculture on biodiversity and abundance: a meta-analysis. *Journal of Applied Ecology 42*, 261-269.
Bergeson, G. B. (1971). Response of muskmelon to fumigation for control of *Meloidogyne incognita* following one year of non host crop. *Plant Disease Reporter 55*, 55-56.
Berkelmans, R., Ferris, H., Tenuta, M., & van Bruggen, A. H. C. (2003). Long-term effects of crop management on trophic levels of nematodes other than plant parasitic nematodes disappear after one year of uniform treatment. *Applied Soil Ecology 23*, 223-235.
Bessey, E. A. (1911). Root-knot and its control. *Bureau of Plant Industry Bulletin 217*. Washington, D. C.: United States Department of Agriculture.
Bilgrami, A. L., & Brey, C. (2005). Potential of predatory nematodes to control plant-parasitic nematodes. In P. S. Grewal, R.- U. Ehlers, & D. I. Shapiro-Ilan (Eds.), *Nematodes as Biocontrol Agents* (pp. 447-464). Oxfordshire, UK: CABI Publishing.
Bilgrami, A. L., Brey, C., & Gaugler, R. (2008). First field release of a predatory nematode, *Mononchoides gaugleri* (Nematoda: Diplogastrida), to control plant-parasitic nematodes. *Nematology 10*, 143-146.
Birat, R. B. S. (1970). Influence of cropping pattern on the population of *Meloidogyne javanica* (Treub, 1885) Chitwood, 1949. *Indian Phytopathology 22*, 391-393.
Blok, W. J., Lamers, J. G., Termorshuizen, A. J., & Bollen, G. J. (2000). Control of soilborne plant pathogens by incorporating fresh organic amendments followed by tarping. *Phytopathology 90,* 253-259.
Boldyrev, M. I., & Borzykh, G. I. (1983). [Plants in the control of nematodes.] *Zashchita Rastenii, Moscow 9*, 30.
Braithwaite, C. W. D. (1974). Effect of crop sequence and fallow on population of *Rotylenchulus reniformis* in fumigated and untreated soil. *Plant Disease Reporter 58*, 259-261.
Bridge, J. (1983). Nematodes. In *Pest Control in Tropical Tomatoes* (pp. 69-84). London, UK: Centre for Overseas Pest Research, Overseas Development Administration.
Bridge, J. (1996). Nematode management in sustainable and subsistence agriculture. *Annual Review of Phytopathology 34*, 201-225.
Britto, D. T., Siddiqi, M. Y., Glass, A. D. M., & Kronzucker, H. J. (2001). Futile transmembrane NH_4^+ *cycling: a cellular hypothesis to explain ammonium toxicity in plants.* Proceedings of the National Academy of Sciences, USA 98, *4255-4258.*
Brodie, B. B., Good, J. M., & Jaworski, C. A. (1970). Population dynamics of plant nematodes in cultivated land: effect of summer cover crops in newly cleared land. *Journal of Nematology 2*, 217-222.
Brodie, B. B., Scurrah, M., & Plaisted, R. L. (2000). Release of germplasm resistant to multiple races of potato cyst nematodes. *American Journal of Potato Research 77*, 207-209.
Brown, D. J. F., Zheng, J., & Zhou, X. (2004). Virus vectors. In Z. X. Chen, S. Y. Chen, & D. W. Dickson (Eds.), *Nematology: Advances and Perspectives: Nematode Management*

and Utilization (Volume 2, pp. 717-770). Cambridge, MA: Tsinghua University Press and CABI Publishing.

Cabanillas, E., & Barker, K. R. (1989). Impact of *Paecilomyces lilacinus* inoculum level and application time on control of *Meloidogyne incognita* on tomato. *Journal of Nematology 21*, 115-120.

Carneiro, R. G., & Carneiro, R. M. D. G. (1982). Seleção preliminary de plantas para rotação de culturas em areas infestados por *M. incognita* nos anos de 1979 e 1980. In *Trabalhos Apresantados a VI Reuniao Brasileira de Nematologia* (no. 6, pp. 141-148), 8-12 Feb 1982, Fortaleza.

Carter, W. W., & Nieto, S. Jr. (1975). Population development of *Meloidogyne incognita* as influenced by crop rotation and fallow. *Plant Disease Reporter 59*, 402-403.

Caubel, G., Curvale, J. P., Hemery, J., & Le Bohec, F. (1981). Observations écologiques et experimentations des méthods de lute contre *Pratylenchus penetrans*, nématode nuisible à la l'artichaut en Bretagne. In *3^0 Congresso Internazionale di Studi sul Carciofo* (pp. 985-1002), 1979, Bari, Italy.

Caubel, G., Hemery, J., & Le Bohec, F. (1978). Observations sur l'effet de *Tagetes patula* L. et *T. erecta* L. sur *Pratylenchus penetrans* Cobb. *Extrait de la Revue La Défense des Végétaux 194*, 1-10.

Chellemi, D. O., Olson, S. M., Mitchell, D. J., Seeker, I., & McSorley, R. (1997). Adaptation of soil solarization to the integrated management of soilborne pests of tomato under humid conditions. *Phytopathology 87,* 250-258.

Chen, J., Abawi, G. S., & Zuckerman, B. M. (1999). Suppression of *Meloidogyne hapla* and its damage to lettuce grown in a mineral soil amended with chitin and biocontrol organisms. *Supplement to the Journal of Nematology 31*, 719-725.

Chen, S., & Dickson, D. W. (2004). Biological control of nematodes by fungal antagonists. In Z. X. Chen, S. Y. Chen, & D. W. Dickson (Eds.), *Nematology: Advances and Perspectives: Nematode Management and Utilization* (Volume 2, pp. 979-1039). Cambridge, MA: Tsinghua University Press and CABI Publishing.

Chen, S. Y., Dickson, D. W., & Mitchell, D. J. (1996). Pathogenicity of fungi to *Heterodera glycines*. *Journal of Nematology 28*, 145-158.

Chen, S. Y., Porter, P. M., Reese, C. D., Klossner, L. D., & Stienstra, W. C. (2001a). Evaluation of soybean and pea as trap crops for managing *Heterodera glycines*. *Journal of Nematology 33*, 214-218.

Chen, S. Y., Porter, P. M., Reese, C. D., & Stienstra, W. C. (2001b). Crop sequence effects on soybean cyst nematode and soybean and corn yields. *Crop Science 41*, 1843-1849.

Chen, Z. X., & Dickson, D. W. (1998). Review of *Pasteuria penetrans*: biology, ecology, and biological control potential. *Journal of Nematology 30*, 313-340.

Chen, Z. X., & Dickson, D. W. (2004). Biological control of nematodes with bacterial antagonists. In Z. X. Chen, S. Y. Chen, & D. W. Dickson (Eds.), *Nematology: Advances and Perspectives: Nematode Management and Utilization* (Volume 2, pp. 1041-1082). Cambridge, MA: Tsinghua University Press and CABI Publishing.

Chet, I. (1990). Biological control of soilborne pathogens with fungal antagonists in combination with soil treatments. In D. Hornby, R. J. Cook, Y. Henis, W. H. Ko, A. D. Rovira, B. Schippers, & P. R. Scott (Eds.), *Biological Control of Soilborne Plant Pathogens* (pp. 15-25). Wallingford, UK: CABI Publishing.

Chikaoka, I. (1983). Studies on damage of crops by the root lesion nematode, *Pratylenchus penetrans* Cobb, and control measures especially by utilization of marigold (*Tagetes* spp.). *Bulletin of Kanagawa Horticulture Experiment Station 125*, 1-72 (Japanese with English summary).

Chitwood, D. J. (2002). Phytochemical based strategies for nematode control. *Annual Review of Phytopathology 40*, 221-249.

Cobb, N. A. (1920). Transfer of nematodes (mononchus) from place to place for economic purposes. *Science 51*, 640-641.

Cohn, E., & Mordechai, M. (1974). Experiments in suppressing citrus nematode populations by use of a marigold and a predacious nematode. *Nematologia Mediterranea 2*, 43-53.

Conn, K. L., & Lazarovits, G. (2000). Soil factors influencing the efficacy of liquid swine manure added to soil to kill *Verticillium dahliae*. *Canadian Journal of Plant Pathology 22,* 400-406.

Conn, K. L., Tenuta, M., & Lazarovits, G. (2005). Liquid swine manure can kill *Verticillium dahliae* microsclerotia in soil by volatile fatty acid, nitrous acid, and/or ammonia toxicity. *Phytopathology 95,* 28-35.

Cooke, D. A. (1985). The effect of resistant cultivars of catch crops on the hatching of *Heterodera schachtii*. *Annals of Applied Biology 106*, 111-120.

Cronin, D., Moënne-Loccoz, Y., Dunne, C., & O'Gara, F. (1997). Inhibition of egg hatch of the potato cyst nematode *Globodera rostochiensis* by chitinase-producing bacteria. *European Journal of Plant Pathology 103*, 433-440.

Crow, W. T. (2005). Alternatives to fenamiphos for management of plant-parasitic nematodes on bermudagrass. *Journal of Nematology 37*, 477-482.

Crow, W. T., Weingartner, D. P., Dickson, D. W., & McSorley, R. (2001). Effect of sorghum-sudangrass and velvetbean cover crops on plant-parasitic nematodes associated with potato production in Florida. *Supplement to the Journal of Nematology 33*, 285-288.

Curto, G., Dallavalle, E., & Lazzeri, L. (2005). Life cycle duration of *Meloidogyne incognita* and host status of Brassicaceae and Capparasceae selected for glucosinolate content. *Nematology 7*, 203-212.

Curto, G., Lazzeri, L., Dallavalle, E., & Santi, R. (2008). Management of *Meloidogyne incognita* (Kofoid *et* White) Chitw. in organic horticulture. *Program and abstracts, Third International Biofumigation Symposium* (p. 45), 21-25 Jul 2008. Canberra, Australia: CSIRO Discovery Centre.

Dababat, A. A., & Sikora, R. A. (2007). Induced resistance by the mutualistic endophyte, *Fusarium oxysporum* 162, toward *Meloidogyne incognita* on tomato, *Biocontrol Science and Technology 17,* 969-975.

D'Addabbo, T. (1995). The nematicidal effect of organic amendments: a review of the literature, 1982-1994. *Nematologia Mediterranea 23*, 299-305.

Daulton, R. A. C. (1963). Controlling *M. javanica* in South Rhodesia. *Rhodesia Agricultural Journal 60*, 150-152.

Debprasad, R., Prasad, D., Singh, R. P., & Ray, D. (2000). Chemical examination and antinemic activity of marigold (*Tagetes erecta* L.) flower. *Annals of Plant Protection Sciences 8*, 212-217.

De Jin, R., Won Suh, J., Dong Park, R., Woong Kim, Y., Krishnan, H. B., & Kim, K. (2005). Effect of chitin compost and broth on biological control of *Meloidogyne incognita* on tomato (*Lycopersicon esculentum* mill.). *Nematology 7,* 125-132.

de Leij, F. A. A. M., & Kerry, B. R. (1991). The nematophagous fungus, *Verticillium chlamydosporium,* as a potential biological control agent for *Meloidogyne arenaria. Revue de Nématolgie 14*, 157-164.

Dickson, D. W., Oostendorp, M., Giblin-Davis, R. M., & Mitchell, D. J. (1994). Control of plant-parasitic nematodes by biological antagonists. In D. Rosen, F. D. Bennett, & J. L. Capinera (Eds.), *Pest Management in the Subtropics, Biological Control—a Florida Perspective* (pp. 575-601). Hampshire, UK: Intercept Ltd.

Di Muro, A. (1975). Root knot nematode (*Meloidogyne incognita*) control by two- and three-year crop rotations. *Annali dell' Istituto Sperimentale per il Tabacco 2*, 89-91.

Djian-Caporalino, C., Fazari, A., Arguel, M., Vernie, T., VandeCasteele, C., Faure, I., Brunoud, G., Pijarowski, L., Palloix, A., Lefebvre, V., & Abad, P. (2007). Root-knot nematode (*Meloidogyne* spp.) *Me* resistance genes in pepper (*Capsicum annuum* L.) are clustered on the P9 chromosome. *Theoretical and Applied Genetics 114*, 473-486.

Docherty, P. A., & Snider, M. D. (1991). Effect of hypertonic and sodium-free medium on transport of a membrane glycoprotein along the secretory pathway in cultured mammalian cells. Journal of Cellular Physiology 146, *34-42.*

Ducusin, A. R., & Davide, R. G. (1972). *Meloidogyne incognita*: its effect on tomato yield and some methods of control. *Philippine Agriculturist 55*, 261-281.

Duplessis, M. C. F., & Kroontje, W. (1964). The relationship between pH and ammonia equilibria in soil. *Proceedings of the Soil Science Society of America 28*, 751-754.

Fallon, D. J., Kaya, H. K., Gaugler, R., & Sipes, B. S. (2002). Effects of entomopathogenic nematodes on *Meloidogyne javanica* in tomatoes and soybeans. *Journal of Nematology 34*, 239-245.

Fallon, D. J., Kaya, H. K., Gaugler, R., & Sipes, B. S. (2004). Effect of *Steinernema feltiae-Xenorhabdus bovienii* insect pathogen complex on *Meloidogyne javanica. Nematology 6*, 671-680.

Fassuliotis, G., & Skucas, G. P. (1969). The effect of pyrrolizidine alkaloid ester and plants containing pyrrolizidine on *Meloidogyne incognita* acrita. *Journal of Nematology 1*, 287-288.

Fauzia, M., Jairajpuri, M. S., & Khan, Z. (1998). Biocontrol potential of *Mononchoides longicaudatus* on *Meloidogyne incognita* on tomato plants. *International Journal of Nematology 8*, 89-91.

Fernandez, C., Pinochet J., Esmenjaud, D., Salesses, G., & Felipe. A. (1994). Resistance among new *Prunus* rootstocks and selections to root-knot nematodes in Spain and France. *HortScience 29*, 1064-1067.

Ferris, H., Venette, R. C., & Lau, S. S. (1996). Dynamics of nematode communities in tomatoes grown in conventional and organic farming systems, and under their impact on soil fertility. *Applied Soil Ecology 3*, 161-175.

Fery, R. L., Dukes, P. D. Sr., & Thies, J. A. (1998). 'Carolina Wonder' and 'Charleston Belle': Southern root-knot nematode resistant bell peppers. *HortScience 33*, 900-902.

Fichtner, E., Koppen, D., & Korschens, M. (1984). Bezeihung zwischen Zuckerrübenanteil in der Fruchtfolge, Ertrag und Populationsdynamik van *Heterodera schachtii* Schm. Auf dem Löss-Schwartzende-Standort Bad Lauchstädt. *Archiv für Acker-und Planzenbau und Bodenkunde 28*, 617-623.

Flancl, L. J., & Dropkin, V. H. (1986). *Heterodera glycines* population dynamics and relation of initial population and soybean yield. *Plant Disease 70*, 791-795.

Fleming, M. R., Janowiak, J. J., Kimmel, J. D., Halbrendt, J. M., Bauer, L. S., Miller, D. L., & Hoover, K. (2005). Efficacy of commercial microwave equipment for eradication of pine wood nematodes and cerambycid larvae infesting red pine. *Forest Products Journal 55*, 226-232.

Galper, S., Cohn, E., Spiegel, Y., & Chet, I. (1990). Nematicidal effect of collagen-amended soil and the influence of protease and collagenase. *Revue de Nématologie 13*, 67-71.

Galper, S., Cohn, E., Spiegel, Y., & Chet, I. (1991). A collagenolytic fungus, *Cunninghamella elegans*, for biological control of plant parasitic nematodes. *Journal of Nematology 23,* 269-274.

Gamliel, A., & Stapleton, J. J. (1993). Characterization of antifungal volatile compounds evolved from solarized soil amended with cabbage residues, *Phytopathology 83*, 899-905.

Gamliel, A., & Stapleton, J. J. (1995). Improved soil disinfestation by biotoxic volatile compounds generated from solarized, organic-amended soil. *Acta Horticulturae 382*, 200-205.

Garabedian, S., & Van Gundy, S. D. (1983). Use of avermectins for the control of *Meloidogyne incognita* on tomato. *Journal of Nematology 15*, 503-510.

Giannakou, I. O., Anastasiadis, I. A., Gowen, S. R., & Prophetou-Athanasiadou, D. A. (2007). Effects of a non-chemical nematicide combined with soil solarization for the control of root-knot nematodes. *Crop Protection 26*, 1644-1654.

Giannakou, I. O., & Prophetou-Athanasiadou, D. A. (2004). A novel non-chemical nematicide for the control of root-knot nematodes, *Applied Soil Ecology 26,* 69-79.

Gnanapragasam, N. C. (1981). The influence of cultivating *Eragrostis curvula* in nematode infested soil, on the development on the subsequent built-up of populations in replanted tea. *Tea Quarterly 50*, 160-162.

Gnanavelrajah, N., & Kumaragamae, D. (1999). Nitrogen leaching losses and plant response to nitrogen fertilizers as influenced by application of Neem (*Azadirachta indica* A. Juss) materials. *Tropical Agricultural Research 11*, 29-41.

Golden, A. M., & Shafer, T. (1959). Host-parasite relationships of various plants and the sugar-beet nematode (*Heterodera schachtii*). *Plant Disease Reporter 43*, 1258-1262.

Golinowski, W., Sobczak, M., Kurek, W., & Grymaszewska, G. (1997). The structure of syncytia. In C. Fenoll, F. M. W. Grunder, & S. A. Ohl (Eds.), *Cellular and Molecular Aspects of Plant-Nematode Interactions* (pp. 80-97). Dordrecht, The Netherlands: Kluwer Academic Publishers.

Gommers, F. J., & Bakker, J. (1988). Physiological diseases induced by plant response or products. In G. O. Poinar, & H.- B. Jansson (Eds.), *Diseases of Nematodes* (Volume 1, pp. 3-22). Boca Raton, FL: CRC Press Inc.

Good, J. M. Jr., Robertson, W. K., & Thompson, L. G. Jr. (1954). Effect of crop rotation on the populations of meadow nematodes, *Pratylenchus leiocephalus*, in Norfolk fine sand. *Plant Disease Reporter 38*, 178-180.

Gotlieb, D., Oka, Y., Ben-Daniel, B.- H., & Cohen, Y. (2003). Dry mycelium of *Penicillium chrysogenum* protects cucumber and tomato plants against the root-knot nematode *Meloidogyne javanica*. *Phytoparasitica 31*, 217-225.

Gray, N. F. (1988). Fungi attacking vermiform nematodes. In G. O. J. Poinar, & H. B. Jansson (Eds.), *Diseases of Nematodes* (Volume 2, pp. 3-38). Boca Raton, FL: CRC Press.

Greco, N., Di Vito, M., Saxena, M. C., & Reddy, M. V. (1988). Investigation on the root lesion nematode *Pratylenchus thornei*, in Syria. *Nematologia Mediterranea 16*, 101-105.

Grewal, P. S., Martin, W. R., Miller, R. W., & Lewis, E. (1997). Suppression of plant-parasitic nematode populations in turfgrass by application of entomopathogenic nematodes. *Biocontrol Science and Technology 7*, 393-399.

Griffith, B. S., Ritz, K., & Wheatley, R. E. (1994). Nematodes as indicators of enhanced microbiological activity in a Scottish organic farming system. *Soil Use and Management10*, 20-24.

Grinstein, A., Kritzman, G., Hetzroni, A., Gamliel, A., Mor, M., & Katan, J. (1995). The border effect of soil solarization. *Crop Protection 14*, 315-320.

Guerrero, M. M., Ros, C., Martinez, M. A., Martinez, M. C., Barcelo, N., Lacasa, A., Ros, C., Guirao, P., Bello, A., & Lopez, J. A. (2005). Biofumigation plus solarization efficacy for soil disinfestation in sweet pepper greenhouse in the southeast of Spain. *Acta Horticulturae 698*, 293-297.

Halbrendt, J. M. (1996). Allelopathy in the management of plant-parasitic nematodes. *Journal of Nematology 28*, 8-14.

Halbrendt, J. M., & Jing, G. (1994). Nematode suppressive rotation crops for orchard renovation. *Acta Horticulturae* 363, 49-56.

Hallmann, J., Frankenberg, A., Paffrath, A., & Schmidt, H. (2007). Occurrence and importance of plant-parasitic nematodes in organic farming in Germany. *Nematology 9*, 869-879.

Hallmann, J., & Sikora, R. A. (1994). Occurrence of plant parasitic nematodes and non-pathogenic species of *Fusarium* in tomato plants in Kenya and their role as mutualistic synergists for biological control of root-knot nematodes. *International Journal of Pest Management 40*, 321-325.

Hallmann, J., & Sikora, R. A. (1996). Toxicity of fungal endophyte secondary metabolites to plant parasitic nematodes and soil-borne plant pathogenic fungi. *European Journal of Plant Pathology 102,* 155-162.

Hewlett, T. E., Gerber, J. F., Smith, K. S., & White, J. H. (2002). *In vitro* culture of *Pasteuria penetrans*. *Nematology 4*,152-153.

Hijink, M. J. (1972). An integrated approach in the control of potato root eelworm. *EPPO Bulletin 7,* 41-48.

Hoitink, H. A. J., Stone, A. G., & Han, D. Y. (1997). Suppression of diseases by composts. *HortScience 32*, 184-187.

Hole, D. G., Perkins, A. J., Wilson, J. D., Alexander, I. H., Grice, P. V., & Evans, A. D. (2005). Does organic farming benefit biodiversity? *Biological Conservation 122*, 113-130.

Hooper, D. J., & Evans, K. (1993). Extraction, identification and control of plant parasitic nematodes. In K. Evans, D. L. Trudgill, & J. M. Webster (Eds.), *Plant Parasitic Nematodes in Temperate Agriculture* (pp. 1-59). Oxon, UK: CABI Publishing.

Hu, K., Li, J., & Webster, J. M. (1995). Mortality of plant-parasitic nematodes caused by bacterial (*Xenorhabdus* spp. and *Photorhabdus luminescens*) culture media. *Journal of Nematology 27*, 502-503.

Hu, K., Li, J., & Webster, J. M. (1999). Nematicidal metabolites produced by *Photorhabdus luminescens* (Enterobacteriaceae), bacterial symbiont of entomopathogenic nematodes. *Nematology 1*, 457-469.

Huang, C. S., Charchar, J. M., & Tenente, R. C. V. (1980). Controle de nematoide das galhas em cenoura atraves de rotacao. *Fitopatologia Brasileira 5*, 329-336.

Hussey, R. S., & Boerma, H. R. (1983). Influence of planting date on damage to soybean caused by *Heterodera glycines*. *Journal of Nematology 15*, 253-258.

Hussey, R. S., & Grundler, F. M. W. (1998). Nematode parasitism of plants. In N. R. Perry, & D. J. Wright (Eds.), *The Physiology and Biochemistry of Free-Living and Plant-Parasitic Nematodes* (pp. 213-243). Wallingford, UK: CABI Publishing.

Ichinohe, M. (1984). Integrated control of the root-knot nematode, *Meloidogyne incognita*, on black pepper plantations in the Amazonian region. *Agriculture, Ecosystem and Environment 12*, 271-283.

Ioannou, N. (2001). Integrating soil solarization with grafting on resistant rootstocks for management of soilborne pathogens of eggplant. *Journal of Horticultural Science and Biotechnology 76*, 396-401.

Ishibashi, N., & Choi, D.- R. (1991). Biological control of soil pests by mixed application of entomopathogenic and fungivorous nematodes. *Journal of Nematology 23*, 175-181.

Isman, M. (1999). Pesticides based on plant essential oils. *Pesticide Outlook 10*, 68-72.

Jaffee, B. A. (2004). Do organic amendments enhance the nematode-trapping fungi *Dactylellina haptotyla* and *Arthrobotrys oligospora*? *Journal of Nematology 36*, 267-275.

Jagdale, G. B., & Grewal, P. S. (2004). Effectiveness of a hot water drench for the control of foliar nematodes *Aphelenchoides fragarie* in floriculture. *Journal of Nematology 36*, 49-53.

Jansson, R. K., & Dybas, R. A. (1998). Avermectins: biochemical mode of action, biological activity and agricultural importance. In I. Ishaaya, & D. Degheele (Eds.), *Insecticides with Novel Mode of Action: Mechanisms and Application* (pp. 152-167). New York: Springer-Verlag.

Jatala, P. (1986). Biological control of plant parasitic nematodes. *Annual Review of Phytopathology 24*, 453-489.

Johnson, A. W., Dowler, C. C., & Hauser, E. W. (1975). Crop rotation and herbicide effects on population densities of plant-parasitic nematodes. *Journal of Nematology 7*, 158-168.

Johnson, A. W., Golden, A. M., Auld, D. L., & Sumner, D. R. (1992). Effects of rapeseed and vetch as green manure crops and fallow on nematodes and soil-borne pathogens. *Journal of Nematology 24*, 117-126.

Jones, M. G. K., & Northcote, D. H. (1972). Multinucleate transfer cells induced in *Coleus* roots by the root-knot nematode, *Meloidogyne arenaria*. *Protoplasma 75*, 381-395.

Jourand, P., Rapior, S., Fargette, M., & Mateille T. (2004). Nematostatic effect of a leaf extract from *Crotalaria virgulata* subsp. *grantiana* on *Meloidogyne incognita* and its use to protect tomato roots. *Nematology 6*, 79-84.

Kaloshian, I., Williamson, V. M., Miyao, G., Lawn, D. A., & Westerdahl, B. B. (1996). "Resistance-breaking" nematodes identified in California tomatoes. *California Agriculture 50*, 18-19.

Kaplan, M., & Noe, J. P. (1993). Effect of chicken-excrement amendments on *Meloidogyne arenaria*. *Journal of Nematology 25*, 71-77.

Katan, J. (1981). Solar heating (solarization) of soil for control of soilborne pests. *Annual Review of Phytopathology 19*, 211-236.

Keren-Zur, M., Antonov, J., Bercovitz, A., Feldman, K., Husid, A., Kenan, G., Markov, N., & Rebhun, M. (2000). *Bacillus firmus* formulations for the safe control of root-knot nematodes. In *Proceedings of the Brighton Crop Protection Conference on Pests and Diseases* (Volume 2A, pp. 47-52). Farnham, UK: British Crop Protection Council.

Kerry, B. R. (1975). Fungi and decrease of cereal cyst-nematode populations in cereal monoculture. *EPPO Bulletin 5*, 353-361.

Kerry, B. R., & Jaffee, B. A. (1997). Fungi as biocontrol agents for plant parasitic nematodes. In D. T. Wicklow, & B. Söderström (Eds.), *The Mycota: a Comprehensive Treatise on Fungi as Experimental System for Basic Applied Research* (Volume 4, pp. 203-218). Berlin, Germany: Springer-Verlag.

Khan, A., Williams, K., & Nevalainen, H. (2006). Control of plant-parasitic nematodes by *Paecilomyces lilacinus* and *Monacrosporium lysipagum* in pot trials. *BioControl 51*, 643-658.

Khan, Z., & Kim, Y. H. (2005). The predatory nematode, *Mononchoides fortidens* (Nematoda: Diplogasterida), suppresses the root-knot nematode, *Meloidogyne arenaria*, in potted field soil. *Biological Control 35*, 78-82.

Kiewnick, S., & Sikora, R. A. (2006). Evaluation of *Paecilomyces lilacinus* strain 251 for the biological control of the northern root-knot nematode *Meloidogyne hapla* Chitowood. *Nematology 8*, 69-78.

Kim, D. G., & Riggs, R. D. (1991). Characteristics and efficacy of a sterile hypomycete (ARF18), a new biocontrol agent for *Heterodera glycines* and other nematodes. *Journal of Nematology 23*, 275-282.

Kimpinski, J., Arsenault, W. J., Gallant, C. E., & Sanderson, J. B. (2000). The effect of marigolds (*Tagetes* spp.) and other cover crops on *Pratylenchus penetrans* and on following potato crops. *Journal of Nematology 32*, 531-536.

Kimpinski, J., Gallant, C. E., Henry, R., Macleod, J. A., Sanderson, J. B., & Sturz, A. V. (2003). Effect of compost and manure soil amendments on nematodes and on yields of potato and barley: a 7-year study. *Journal of Nematology 35*, 289-293.

Kirkegaard, J. A., & Sarwar, M. (1998). Biofumigation potential of brassicas. *Plant and Soil 201*, 71-89.

Kiryanova, E. S., & Krall, E. L. (1971). *Plant Parasitic Nematodes and Their Control* (Volume 2). Leningrad: Nauka Publisher (English translation: Amerind Publishing, New Delhi).

Kloepper, J. W., Rodríguez-Kábana, R., McInroy, J. A., & Young, R. W. (1992). Rhizosphere bacteria antagonistic to soybean cyst (*Heterodera glycines*) and root-knot (*Meloidogyne incognita*) nematodes: identification by fatty acid analysis and frequency of biological control activity. *Plant Soil 139*, 75-84.

Kluepfel, D. A., Nyczepir, A. P., Lawrence, J. E., Wechter, W. P., & Leverentz, B. (2002). Biological control of the phytoparasitic nematode *Mesocriconema xenoplax* on peach trees. *Journal of Nematology 34*, 120-123.

Koch, D. W., & Gray, F. A. (1997). Nematode-resistant oil radish for control of *Heterodera schachtii*. I. Sugar beet-barley rotations. *Journal of Sugar Beet Research 34*, 31-43.

Kokalis-Burelle, N., Vavrina, C. S., Reddy, M. S., & Kloepper, J. W. (2003). Amendment of muskmelon transplant media with plant growth-promoting rhizobacteria: effects on seedling quality, disease, and nematode resistance. *Hortechnology 13*, 476-482.

Kokalis-Burelle, N., Vavrina, C. S., Rossskopf, E. N., & Shelby, R. A. (2002). Field evaluation of plant growth-promoting rhizobacteria amended transplant mixes and soil solarization for tomato and pepper production in Florida. *Plant and Soil 238*, 257-266.

Krishnaveni, M., & Subramanian, S. (2005). Root-knot nematodes of cucurbits and their management—a review. *Agricultural Reviews—Agricultural Research Communications Centre India 26*, 103-113.

Lamers, J. G., & Roosjen, J. (1993). De waardplantgeschikthikheid van groenbemestings gewassen voor het NoordelijK wortl-knobbelaaltje. *Verlag- Proefsstation voor de Akkerbouw en de Groenreteelt in de Vollegrond No. 163*.

LaMondia, J. A. (2006). Management of lesion nematodes and potato early dying with rotation crops. *Journal of Nematology 38*, 442-448.

LaMondia, J. A., & Brodie, B. B. (1984). Control of *Globodera rostochiensis* by solar heat. *Plant Disease 68*, 474-476.

LaMondia, J. A., Gent, M. P. N., Ferrandino, F. J., Elmer, W. H., & Stoner, K. A. (1999). Effect of compost amendment of straw mulch on potato early dying disease. *Plant Disease 83*, 361-366.

Lazarovits, G., Tenuta, M., & Conn, K. L. (2000). Utilization of high nitrogen and swine manure amendments for control of soil-borne diseases: efficacy and mode of action. *Acta Horticulturae 532*, 59-64.

Lazzeri, L., Curto, G., Leoni, O., & Dallavalle, E. (2004a). Effects of glucosinolates and their enzymatic hydrolysis products via myrosinase on the root-knot nematode *Meloidogyne incognita* (Kofoid et White) Chitw. *Journal of Agriculture, Food and Chemistry 52*, 6703-6707.

Lazzeri, L., Leoni, O., Bernardi, R., Malaguti, L., & Cinti, S. (2004b). Plants, techniques and products for optimizing biofumigation in full field. *Agroindustria 3*, 281-288.

Lazzeri, L., Leoni, O., & Manici, L. M. (2004c). Biocidal plant dried pellets for biofumigation. *Industrial Crops and Products 20*, 59-65.

Lazzeri, L., Tacconi, R., & Palmieri, S. (1993). In vitro activity of some glucosinolates and their reaction products toward a population of the nematode *Heterodera schachtii*. *Journal of Agriculture, Food and Chemistry 41*, 825-829.

Lewis, S. A., & Fassuliotis, G. (1982). Lance nematodes, *Hoplolaimus* spp., in the southern region of the United States. In *South Cooperative Service Bulletin 276* (pp. 127-138). Fayetteville, AR: University of Arkansas.

Linford, M. B. (1937). Stimulated activity of natural enemies of nematodes. *Science 85*, 123-124.

Linford, M. B., Yap, F., & Oliveira, J. M. (1938). Reduction of soil populations of root-knot nematode during decomposition of organic matter. *Soil Science 45*, 127-141.

Liu, S., & Chen, S. (2005). Efficacy of the fungi *Hirsutella minnesotensis* and *H. rhossiliensis* from liquid culture for control of the soybean cyst nematode *Heterodera glycines*. *Nematology 7*, 149-157.

MacGuidwin, A. E., & Layne, T. L. (1995). Response of nematode communities to sudangrass and sorghum-sudangrass hybrids grown as green manure crops. *Journal of Nematology 27*, 609-616.

Madulu, J. D., Trudgill, D. L., & Philipps, M. S. (1994). Rotational management of *Meloidogyne javanica* and effects on *Pasteuria penetrans* and tomato and tobacco yields. *Nematologia 40*, 438-455.

Mahran, A., Tenuta, M., Hanson, M., & Daayf, F. (2008). Mortality of *Pratylenchus penetrans* by volatile fatty acids from liquid hog manure. *Journal of Nematology 40*, 119-126.

Manzanilla-Lopez, R. H., Evans, K., & Bridge, J. (2004). Plant diseases caused by nematodes. In Z. X. Chen, S. Y. Chen, & D. W. Dickson (Eds.), *Nematology: Advances and Perspectives: Nematode Management and Utilization* (Volume 2, pp. 637-716). Cambridge, MA: Tsinghua University Press and CABI Publishing.

Martin, M. J., Riedel, R. M., & Rowe, R. C. (1982). *Verticillium dahlia* and *Pratylenchus penetrans*: interactions in the early dying complex of potato in Ohio. *Phytopathology 72*, 640-644.

Marull, J., Pinochet, J., & Rodríguez-Kábana, R. (1997). Agricultural and municipal compost residues for control of root-knot nematodes in tomato and pepper. *Compost Science and Utilization 1*, 6-15.

Matthiessen, J. N., & Kirkegaard, J. A. (2006). Biofumigation and enhanced biodegradation: opportunity and challenge in soilborne pest and disease management. *Critical Reviews in Plant Sciences 25*, 235-265.

Mavrogianopoulos, G. N., Frangoudakis, A., & Pandelakis, J. (2000). Energy efficient soil disinfestation by microwaves. *Journal of Agricultural Engineering Research 75*, 149-153.

McLeod, R. W., & Steel, C. C. (1999). Effects of brassica-leaf green manures and crops on activity and reproduction of *Meloidogyne javanica*. *Nematology 1*, 613-624.

McSorley, R. (1999). Host suitability of potential cover crops for root-knot nematodes. *Journal of Nematology 31(4S)*, 619-623.

McSorley, R., & Dickson, D. W. (1995). Effect of tropical rotation crops on *Meloidogyne incognita* and other plant-parasitic nematodes. *Journal of Nematology 27*, 535-544.

McSorley, R., Dickson, D. W., de Brito, J. A., Hewlett, T. E., & Frederick, J. J. (1994). Effect of tropical rotation crops on *Meloidogyne arenaria* population densities and vegetable yields in microplots. *Journal of Nematology 26*, 175-181.

McSorley, R., Dickson, D. W., de Brito, J. A., & Hochmuth, R. C. (1994). Tropical rotation crops influence nematode densities and vegetable yields. *Journal of Nematology 26*, 308-314.

McSorley, R., & Gallaher, R. N. (1995). Effect of yard waste compost on plant-parasitic nematode densities in vegetable crops. *Journal of Nematology 27(4S)*, 545-549.

McSorley, R., & Gallaher, R. N. (1996). Effect of yard waste compost on nematode densities and maize yield. *Journal of Nematology 28(4S),* 655-660.

Meagher, J. W., & Rooney, D. R. (1966). The effect of crop rotations in the Victorian Wimmera on the cereal cyst nematode (*Heterodera avenae*), nitrogen fertility and wheat yield. *Australian Journal of Experimental Agriculture and Animal Husbandry 6*, 425-431.

Mekete, T., Hallmann, J., Kiewnick, S., & Sikora, R. (2009). Endophytic bacteria from Ethiopian coffee plants and their potential to antagonise *Meloidogyne incognita*. *Nematology 11*, 117-127.

Melakeberhan, H., Mennan, S., Ngouajio, M., & Dudek, T. (2008). Effect of *Meloidogyne hapla* on multi-purpose use of oilseed radish (*Raphanus sativus*). *Nematology 10*, 375-379.

Melakeberhan, H., Xu, A., Kravchenko, A., Mennan, S., & Riga, E. (2006). Potential use of arugula (*Eruca sativa* L.) as a trap crop for *Meloidogyne hapla*. *Nematology 8*, 793-799.

Mian, H. I., Godoy, G., Shelby, R., Rodríguez-Kábana, R., & Morgan-Jones, G. (1982). Chitin amendments for control of *Meloidogyne arenarea* in infested soil. *Nematropica 12*, 71-84.

Min, Y. Y., Sato, E., Shirakashi, T., Wada, S., Toyota, K., & Watanabe, A. (2007). Suppressive effect of anaerobically digested slurry on the root lesion nematode *Pratylenchus penetrans* and its potential mechanisms. *Japanese Journal of Nematology 37*, 93-100.

Minnis, S. T., Haydock, P. P. J., Ibrahim, S. K., Grove, I. G., Evans, K., & Russell, M. D. (2002). Potato cyst nematodes in England and Wales: occurrence and distribution. *Annals of Applied Biology 140*, 187-195.

Minton, N. A., & Sayer, R. M. (1989). Suppressive influence of *Pasteuria penetrans* in Georgia soil on reproduction of *Meloidogyne arenaria*. *Journal of Nematology, 21*, 574-575.

Mitchell, B. L., Blair, B. W., & Martin, G. C. (1971). Potatoes in Rhodesia. Part II. Pests of potatoes. *Rhodesia Agricultural Journal Technical Bulletin No. 11,* 25-31.

Mitkowski, N. A., & Abawi, G. S. (2002). Soil organic matter and management of plant-parasitic nematodes. *Journal of Nematology 34*, 289-295.

Mojtahedi, H., Santo, G. S., Hang, A. N., & Wilson, J. H. (1991). Suppression of root-knot nematode populations with selected rapeseed cultivars as green manure. *Journal of Nematology 23*, 170-174.

Mojtahedi, H., Santo, G. S., & Ingham, R. E. (1993). Suppression of *Meloidogyne chitwoodi* with sudangrass cultivars as green manure. *Journal of Nematology 25*, 303-311.

Mojumdar, V. (1995). Effects on nematodes. In H. Schmutterer (Ed.), *The Neem Tree,* Azadirachta indica *A. Juss. and other Meliaceous Plants: Source of Unique Natural Products for Integrated Pest Management, Industry, and other Purposes* (pp. 129-150). Weinheim, Germany: VCH.

Momma, N., Usami, T., Amemiya, A., & Shishido, M. (2005). Factors involved in the suppression of *Fusarium oxysporum* f. sp. *lycopersici* by soil reduction. *Soil Microorgamisms 59,* 27-33.

Momma, N., Yamamoto, K., Simandi, P., & Shishido, M. (2006). Role of organic acids in the mechanisms of biological soil disinfestation (BSD). *Journal of General Plant Pathology 72*, 247-252.

Mor, M., Cohn, E., & Spiegel, Y. (1992). Phenology, pathogenicity and pathotypes of cereal cyst nematodes, *Heterodera avenae* and *H. latipons* (Nematoda: Heteroderidae) in Israel. *Nematologica 38*, 494-501.

Morgan-Jones, G., & Rodríguez-Kábana, R. (1988). Fungi colonizing cysts and eggs. In G. O. J. Poinar, & H. B. Jansson (Eds.), *Diseases of Nematodes* (Volume 2, pp. 39-58). Boca Raton, FL: CRC Press.

Motsinger, R. E., Moody, E. H., & Gay, C. M. (1977). Reaction of certain French marigold (*Tagetes patula*) cultivars to three *Meloidogyne* spp. *Journal of Nematology 9*, 278.

Müller, J. (1991). Catch cropping for population control of *Heterodera schachtii*. In *Proceedings of the 54th Winter Congress, International Institute for Sugar Beet Research* (pp 179-196).

Murphy, W. S., Brodie, B. B., & Good, J. M. (1974). Population dynamics of plant nematodes in cultivated soil: effects of combinations of cropping systems and nematicides. *Journal of Nematology, 6*, 103-107.

Nakasono, K. (1973). [Relationship between cultivation of *Tagetes* plants and changes in population of the reniform nematode, *Rotylenchulus reniformis*.] *Japanese Journal of Nematology 3*, 38-41.

Natarajan, N., Cork, A., Boomathi, N., Pandi, R., Velavan, S., & Dhakshnamoorthy, G. (2006). Cold aqueous extracts of African marigold, *Tagetes erecta* for control tomato root knot nematode, *Meloidogyne incognita*. *Crop Protection 25*, 1210-1213.

Neher, D. A. (1999). Nematode communities in organically and conventionally managed agricultural soils. *Journal of Nematology 31*, 142-154.

Neher, D. A. (2001). Role of nematodes in soil health and their use as indicators. *Journal of Nematology 33*, 161-168.

Netscher, C. (1974). L'arachide et le controle biologique des nématodes *Meloidogyne* spp. dans les cultures maraîchères du Sénegal. *Comptes Rendus des Seances de l'Académie d'Agriculture de France 60*, 1332-1339.

Ngundo, B. W. (1979). Crop rotation to control *Meloidogyne hapla* on pyrethrum. In *IX International Congress of Plant Protection* (p. 132), Aug 1979, Washington, D.C.

Oda, M. (1999). Grafting of vegetables to improve greenhouse production. In *Extension Bulletin* (December). Osaka, Japan: College of Agriculture, Osaka Prefecture University.

Oka, Y., Ben-Daniel, B.- H., & Cohen, Y. (2001). Nematicidal activity of powder and extracts of *Inula viscosa*. *Nematology 3*, 735-742.

Oka, Y., Ben-Daniel B.- H., & Cohen, Y. (2006a). Control of *Meloidogyne javanica* by formulations of *Inula viscosa* leaf extracts. *Journal of Nematology 38*, 46-51.

Oka, Y., Chet, I., Mor, M., & Spiegel, Y. (1997). A fungal parasite of *Meloidogyne javanica* eggs: evaluation of its use to control the root-knot nematode. *Biocontrol Science and Technology 7*, 489-497.

Oka, Y., Chet, I., & Spiegel, I. (1993). Control of the root-knot nematode *Meloidogyne javanica* by *Bacillus cereus*. *Biocontrol Science and Technology 3*, 115-126.

Oka, Y., Koltai, H., Bar-Eyal, M., Mor, M., Sharon, E., Chet, I., & Spiegel, Y. (2000a). New strategies for control of plant-parasitic nematodes. *Pest Management Science 56*, 983-988.

Oka, Y., Nacar, S., Putievsky, E., Ravid, U., Yaniv, Z., & Spiegel, Y. (2000b). Nematicidal activity of essential oils and their components against the root-knot nematode. *Phytopathology* 90, 710-715.

Oka, Y., Offenbach, R., & Pivonia, S. (2004). Pepper rootstock graft compatibility and response to Meloidogyne javanica *and* M. incognita. Journal of Nematology 36, *137-141.*

Oka, Y., & Pivonia, S. (2002). Use of ammonia-releasing compounds for control of the root-knot nematode Meloidogyne javanica. Nematology 4, *65-71.*

Oka, Y., & Pivonia, S. (2003). Effect of a nitrification inhibitor on nematicidal activity of organic and inorganic ammonia-releasing compounds against the root-knot nematode *Meloidogyne javanica. Nematology 5*, 505-513.

Oka, Y., Shapira, N., & Fine, P. (2007a). Control of root-knot nematodes in organic farming system by organic amendments and soil solarization. *Crop Protection 26*, 1556-1565.

Oka, Y., Tkachi, N., Shuker, S., Rosenberg, R., Suriano, S., & Fine, P. (2006b). Field studies on the enhancement of nematicidal activity of ammonia-releasing fertilisers by alkaline amendments. *Nematology 8*, 881-893.

Oka, Y., Tkachi, N., Shuker, S., Rosenberg, R., Suriano, S., & Fine, P. (2006c). Laboratory studies on the enhancement of nematicidal activity of ammonia-releasing fertilizers by alkaline amendments. *Nematology 8*, 335-346.

Oka, Y., Tkachi, N., Shuker, S., & Yerumiyahu, U. (2007b). Enhanced nematicidal activity of organic and inorganic ammonia-releasing amendments using neem extracts. *Journal of Nematology 39*, 9-16.

Oka, Y., & Yermiyahu, U. (2002). Nematode-suppressive effects of composts against the root-knot nematode *Meloidogyne javanica* on tomato. *Nematology 4*, 891-898.

Olatinwo, R., Becker, J. O., & Borneman, J. (2006). Suppression of *Heterodera schachtii* populations by *Dactylella oviparasitica* in four soils. *Journal of Nematology 38*, 345-348.

Oostenbrink, M., Kuiper, K., & S'Jacob, J. J. (1957). *Tagetes* als feindpflanzen von *Pratylenchus* arten. *Nematologia 2* (Suppl.), 434-433.

Osman, G. Y., & Salem, F. M. (1995). Bio-efficacy of sincocin AGTM to control *Tylenchulus semipenetrans* (Tylenchida, Nematoda) in citrus orchard. *Journal of Pest Science 68*, 179-181.

Park, I. K., Park, J. Y., Kim, K. H., Choi, K. S., Choi, I. H., Kim, C. S., & Shin, S. C. (2005). Nematicidal activity of plant essential oils and components from garlic (*Allium sativum*) and cinnamon (*Cinnamomum verum*) oils against the pine wood nematode (*Bursaphelenchus xylophilus*). *Nematology 7*, 767-774.

Peng, Y., & Moens, M. (2003). Host resistance and tolerance to migratory plant-parasitic nematodes. *Nematology 5*, 145-177.

Ploeg, A. T. (2000). Effects of amending soil with *Tagetes patula* cv. Single Gold on *Meloidogyne incognita* infestation of tomato. *Nematology 2*, 489-493.

Ploeg, A. T. (2002). Effects of selected marigold varieties on root-knot nematodes and tomato and melon yields. *Plant Disease 86*, 505-508.

Ploeg, A. T., & Stapleton, J. J. (2001). Glasshouse studies on the effects of time, temperature and amendment of soil with broccoli plant residues on the infestation of melon plants by *Meloidogyne incognita* and *M. javanica*. *Nematology 3*, 855-861.

Potter, M. J., Vanstone, V. A., Davies, K. A., Kirkegaard, J. A., & Rathjen, A. J. (1999). Reduced susceptibility of *Brassica napus* to *Pratylenchus neglectus* in plants with elevated root levels of 2-phenylethyl glucosinolate. *Journal of Nematology 31*, 291-298.

Prasad, D., Nagla, D. K., Kumar, S., & Saini, M. L. (1992). Marigold plants for management of nematode populations in fields. *Current Nematology 3*, 15-18.

Radwan, M. A., El-Maadawy, E. K., Kassem, S. I., & Abu-Elamayem, M. M. (2009). Oil cakes soil amendment effects on *Meloidogyne incognita*, root-knot nematode infecting tomato. *Archives of Phytopathology and Plant Protection 42*, 58-64.

Raguraman, S., Ganapathy, N., & Venkatesan, T. (2004). Neem versus entomopathogens and natural enemies of crop pests: the potential impact and strategies. In O. Koul & S. Wahab (Eds.), *Neem: Today and in the New Millennium* (pp. 125-182). Dordrecht, The Netherlands: Kluwer Academic Publishers.

Rahman, M. L. (1990). Effect of different cropping sequences on root-knot nematode, *Meloidogyne graminicola*, and yield of deepwater rice. *Nematologia Mediterranea 18*, 213-217.

Raviv, M., Oka, Y., Katan, J., Hadar, Y., Yogev, A., Medina, S., Krasnovsky, A., & Ziadan, H. (2005). High-nitrogen compost as a medium for organic container-grown crops. *Bioresource Techno*logy *96*, 419-427.

Reitz, M., Rudolph, K., Schröder, I., Hoffmann-Hergarten, S., Hallmann, J., & Sikora, R. A. (2000). Lipopolysaccharides of *Rhizobium etli* strain G12 act in potato roots as an inducing agent of systemic resistance to infection by the cyst nematode *Globodera pallida*. *Applied and Environmental Microbiology 66*, 3515-3518.

Reshmi, B., & Vijayalakshmi, M. (1998). Effect of aqueous extracts of neem based pesticides on the mortality of second-stage juveniles of *Meloidogyne incognita*. *Annals of Plant Protection Sciences 6*, 53-55.

Reynolds, L. B., Potter, J. W., & Ball Coelho, B. R. (2000). Crop rotation with *Tagetes* sp. is an alternative to chemical fumigation for control of root-lesion nematodes. *Agronomy Journal 92*, 957-966.

Rhoades, H. L. (1976). Effect of *Indigofera hirsute* on *Belonolaimus longicaudatus*, *Meloidogyne incognita* and *M. javanica* and subsequent crop yields. *Plant Disease Reporter 60*, 384-386.

Rhoades, H. L., & Forbes, R. B. (1986). Effect of fallow, cover crops, organic mulches, and fenamiphos on nematode populations, soil nutrients and subsequent crop growth. *Nematropica 16*, 141-151.

Rhode, R. A., & Jenkins, W. R. (1957). Host range of species of *Trichodorus* and its host-parasite relationships on tomato. *Phytopathology 47*, 295-298.

Rich, J. R.., Dunn, R. A., & Noling, J. W. (2004). Nematicides: past and present uses. In Z. X. Chen, S. Y. Chen, & D. W. Dickson (Eds.), *Nematology: Advances and Perspectives: Nematode Management and Utilization* (Volume 2, pp. 1041-1082). Cambridge, MA: Tsinghua University Press and CABI Publishing.

Riegel, C., & Noe, J. P. (2000). Chicken litter soil amendment effects on soilborne microbes and *Meloidogyne incognita* on cotton. *Plant Disease, 84*, 1275-1281.

Riggs, R. D., & Schmitt, D. P. (1988). Complete characterization of the race scheme for *Heterodera glycines*. *Journal of Nematology, 20*, 392-395.

Roberts, P. A. (1992). Current status of the availability, development and use of host plant resistance to nematodes. *Journal of Nematology 24*, 213-227.

Roberts, P. A. (2002). Concepts and consequences of resistance. In J. L. Starr, R. Cook, & J. Bridge (Eds.), *Plant Resistance to Parasitic Nematodes*. Oxon, UK: CABI Publishing.

Rodríguez-Kábana, R. (1986). Organic and inorganic amendments to soil as nematode suppressants. *Journal of Nematology 18*, 129-135.

Rodríguez-Kábana, R., & Ivey, H. (1986). Crop rotation system for the management of *Meloidogyne arenaria* in peanut. *Nematropica 16*, 53-63.

Rodríguez-Kábana, R., King, P. S., & Pope, M. H. (1981). Combinations of anhydrous ammonia and ethylene dibromide for control of nematodes parasitic on soybeans. *Nematropica 11*, 27-41.

Rodríguez-Kábana, R., King, P. S., Robertson, D. G., & Weaver, C. F. (1988a). Potential of crops uncommon to Alabama for management of root-knot and soybean cyst nematodes. *Annals of Applied Nematology 2,* 116-120.

Rodríguez-Kábana, R., King, P. S., Robertson, D. G., Weaver, C. F., & Carden, E. L. (1988b). New crops with potential for management of soybean nematodes. *Nematropica 18*, 45-52.

Rodríguez-Kábana, R., Kokallis-Burelle, N., Robertson, D. G., King, P. S., & Well, L. (1994). Rotation with coastal Bermuda grass, cotton, and Bahia grass for management of *Meloidogyne arenaria* and Southern blight in peanut. *Journal of Nematology 26*, 665-668.

Rodríguez-Kábana, R., Morgan-Jones, G., & Chet, I. (1987). Biological control of nematodes: soil amendments and microbial antagonists. *Plant and Soil 100*, 237-247.

Rodríguez-Kábana, R., Pinochet, J., Robertson, D. G., & Wells, L. (1992). Crop rotation studies with velvetbean *(Mucuna deeringiana)* for the management of *Meloidogyne* spp. *Journal of Nematology 24*, 662-668.

Rodríguez-Kábana, R., Robertson, D. G., King, P. S., & Well, L. (1991a). American jointvetch and partridge pea for the management of *Meloidogyne arenaria* in peanut. *Nematropica 21*, 97-103.

Rodríguez-Kábana, R., Robertson, D. G., Well, L., & King, P. S. (1991b). Soybean-peanut rotation for the management of *Meloidogyne arenaria* and *Sclerotium rolfsii*. *Nematropica 21*, 147-154.

Rodríguez-Kábana, R., Weaver, D. B., Robertson, D. G., Carden, E. L., & Pegues, M. L. (1991c). Additional studies on the use of bahiagrass for the management of root-knot and cyst nematodes in soybean. *Nematropica 21*, 203-210.

Rodríguez-Kábana, R., Weaver, D. B., Robertson, D. G., Young, R. W., & Carden, E. L. (1990a). Sorghum in rotation with soybean for the management of cyst- and root-knot nematodes. *Nematropica 20*, 111-119.

Rodríguez-Kábana, R., Weaver, D. B., Robertson, D. G., Young, R. W., & Carden, E. L. (1990b). Rotation of soybean with two tropical legumes for the management of nematode problems. *Nematropica 20*, 101-110.

Romero, M. D., Valdeslivas, A., Lacasta, C., & Duce, A. (1991). Evolucion de las polaciones de *Heterodera avenae* (Nematoda) y su efecto en el disarrolo y rendimiento de trigo en rotacion y monocultiva. *Suelo y Planta 1*, 323-334.

Roy, A. K. (1979). Effectiveness of rotation with non-host or fallow in reducing infestation of *Meloidogyne graminicola*. *Indian Journal of Nematology 8*, 156-158.

Roy, A. K. (1987). Rice diseases in Assam. *Indian Farming 37*, 25-27.

Rubio-Cabetas, M. J., Lecouls, A. C., Esmenjaud, D., & Salesses, G. (2000). Genetic control for resistance to root-knot nematodes in *Prunus* rootstocks. *Acta Horticulturae 522*, 155-164.

Runia, W. T. (1983). A recent development in steam sterilization. *Acta Horticulturae 152*, 195-200.

Rush, C. M., & Lyda, S. D. (1982). Effects of anhydrous ammonia on mycelium and sclerotia of Phymatotrichum omnivorum. Phytopathology 72, *1085-1089.*

Sahrawat, K. L., & Mukerjee, S. K. (1977). Nitrification inhibitors. I. Studies with karanjin, a furanolflavonoid from karanja (*Pongamia glagra*) seeds. *Plant and Soil 47*, 27-36.

Saka, V. W. (1982). International Meloidogyne Project Report in Malawi. In *Proceedings of the 3rd Research Planning Conference on Root-knot Nematodes,* Meloidogyne *spp. Regions IV and V* (pp. 31-36), 16-20 Nov, Ibadan, Nigeria.

Samaliev, H. Y., Andreoglou, F. I., Elawad, S. A., & Hague, N. G. M. (2000). The nematicidal effects of the bacteria *Pseudomonas oryzihabitans* and *Xenorhabdus nematophilus* on the root-knot nematode *Meloidogyne javanica*. *Nematology 2*, 507-514.

Sànchez de Viala, S., Brodie, B. B., Rodriguez, E., & Gibson, D. M. (1998). The potential of thiarubrine C as a nematicidal agent against plant parasitic nematodes. *Journal of Nematology 30*, 192-200.

Sangwan, N. K., Verma, B. S., Verma, K. K., & Dhindsa, K. S. (1990). Nematicidal activity of some essential plant oils. *Pesticide Science 28,* 331-335.

San Martin, R., & Magunacelaya, J. C. (2005). Control of plant-parasitic nematodes with extracts of *Quillaja saponaria*. *Nematology 7*, 577-585.

Sasser, J. N., & Freckman, D. W. (1987). A world perspective of nematology: the role of the society. In J. A. Veech, & D. W. Dickson (Eds.), *Vistas on Nematology* (pp. 7-14). Hyattsville, MD: Society of Nematologists.

Sasser, J. N., Kirkpatrick, T. L., & Dybas, R. A. (1982). Efficacy of avermectins for root-knot control in tobacco *Meloidogyne incognita*. *Plant Disease 66*, 691-693.

Sayre, R. M., Patrick, Z. A., & Thorpe, H. J. (1965). Identification of a selective nematicidal component in extracts of plant residues decomposing in soil. *Nematologica 11*, 263-268.

Scholte, K. (2000). Growth and development of plants with potential for use as trap crops for potato cyst nematodes and their effects on the numbers of juveniles in cysts. *Annals of Applied Biology 137*, 31-42.

Shaaya, E., Ravid, U., Paster, N., Juven, B., Zisman, U., & Pisarev, V. (1991). Fumigant toxicity of essential oils against four major stored-product insects. *Journal of Chemical Ecology 17*, 499-504.

Shagalina, L. M. (1979). [Use of Lucerne in crop rotation of cotton fields infected with *Meloidogyne acrita*.] *Izvestiya Akademii nauk Turkmenskoi SSR, Biologicheskie Nauki No. 2*, 91-92.

Sharma, S. K., Inderjit, S., & Sakhuja, P. K. (1980). Influence of different cropping sequences on the population of root-knot nematode, *Meloidogyne incognita* and the performance of the subsequent mung crop. *Indian Journal of Nematology 10*, 53-58.

Sharon, E., Bar-Eyal, M., Chet, I., Herrera-Estrella, A., Kleifeld, O., & Spiegel, Y. (2001). Biological control of the root-knot nematode *Meloidogyne javanica* by *Tricoderma harzianum*. *Phytopathology 91*, 687-693.

Shurtleff, M. C., & Averre, C. W. (2000). *Diagnosing Plant Diseases Caused by Nematodes* (p. 187). St. Paul, MN: The American Phytopathological Society.

Sigüenza, C., Schochow, M., Turini, T., & Ploeg, A. (2005). Use of *Cucumis metuliferus* as a rootstock for melon to manage *Meloidogyne incognita*. *Journal of Nematology 37*, 276-280.

Sikora, R. A. (1992). Management of the antagonistic potential in agricultural ecosystems for the biological control of plant parasitic nematodes. *Annual Review of Phytopathology 30*, 245-270.

Sikora, R. A., Pocasangre, L., zum Felde, A., Niere, B., Vu, T. T., & Dababat, A. A. (2008). Mutualistic endophytic fungi and *in-planta* suppressiveness to plant parasitic nematodes. *Biological Control 46*, 15-23.

Sivapalan, P. (1972). Nematode pest of tea. In J. M. Webster (Ed.), *Economic Nematology* (pp. 285-310). London, New York: Academic Press.

Slack, D. A., Riggs, R. D., & Hamblen, M. L. (1972). The effect of temperature and moisture on the survival of *Heterodera glycines* in the absence of a host. *Journal of Nematology 4*, 263-266.

Spiegel, Y., & Chet, I. (1998). Evaluation of *Trichoderma* spp. as a biocontrol agent against soilborne fungi and plant-parasitic nematodes in Israel. *Integrated Pest Management Review 3*, 169-175.

Spiegel. Y., Chet, I., & Cohen, E. (1987). Use of chitin for controlling plant parasitic nematodes. II. Mode of action. *Plant and Soil 98*, 337-345.

Spiegel, Y., Cohn, E., Galper, S., Sharon, E., & Chet, I. (1991). Evaluation of a newly isolated bacterium, *Pseudomonas chitinolytica* sp. nov., for controlling the root-knot nematode *Meloidogyne javanica*. *Biocontrol Science and Technology 1,* 115-125.

Stapleton, J. J., & Duncan, R. A. (1998). Soil disinfestation with cruciferous amendments and sublethal heating: effects on *Meloidogyne incognita, Sclerotium rolfsii* and *Pythium ultimum*. *Plant Pathology 47*, 737-742.

Starr, J. L., Cook, R., & Bridge, J. (2002). *Plant Resistance to Parasitic Nematodes*. Wallingford, UK: CABI Publishing.

Starr, J. L, & Roberts, P. A. (2004). Resistance to plant-parasitic nematodes. In Z. X. Chen, S. Y. Chen, & D. W. Dickson (Eds.), *Nematology: Advances and Perspectives: Nematode Management and Utilization* (Volume 2, pp. 879-907). Cambridge, MA: Tsinghua University Press and CABI Publishing.

Steele, A. E., & Price, C. (1965). The effect of five legume crops on soil populations of the sugar beet nematode (*Heterodera schachtii* Schmidt). *Journal of the American Society of Sugar Beet Technologists 13*, 314-317.

Stirling, G. R. (1984). Biological control of *Meloidogyne javanica* with *Bacillus penetrans*. *Phytopathology 74*, 55-60.

Stirling, G. R. (1991). *Biological Control of Plant Parasitic Nematodes*. Wallingford, UK: CABI Publications.

Stirling, G. R., McKenry, M. V., & Mankau, R. (1979). Biological control of root-knot nematodes (*Meloidogyne* spp.) on peaches. *Phytopathology 69*, 806-809.

Stirling, G. R., & Smith, L. J. (1998). Tests of formulated products containing either *Verticillium chlamydosporium* or *Arthrobotrys dactyloides* for biological control of root-knot nematodes. *Biological Control 3*, 231-239.

Stirling, G. R., & Stirling, A. M. (2003). The potential of *Brassica* green manure crops for controlling root-knot nematode (*Meloidogyne javanica*) on horticultural crops in a subtropical environment. *Australian Journal of Experimental Agriculture 43*, 623-630.

Stirling, G. R., & White, A. M. (1982). Distribution of a parasite of root-knot nematodes in South Australian vineyards. *Plant Disease 66*, 52-53.

Stover, R. H. (1979). Flooding of soil for disease control. In D. Mulder (Ed.), *Soil Disinfestation* (pp. 19-28). Amsterdam, The Netherlands: Elsevier Scientific Publications.

Sundararaj, P., & Mehta, U. K. (1990). Host status of some economic crops to *Pratylenchus zeae* and their influence on subsequent sugarcane crops. *Indian Journal of Nematology 20*, 165-169.

Suresh, G., Gopalakrishnan, G., & Masilamani, S. (2004). Neem for plant pathogenic fungal control: the outlook in the new millennium. In O. Koul, & S. Wahab (Eds.), *Neem: Today and in the New Millennium* (pp. 183-208). Dordrecht, The Netherlands: Kluwer Academic Publishers.

Taba, S., Sawada, J., & Moromizato, Z. (2008). Nematicidal activity of Okinawa Island plants on the root-knot nematode *Meloidogyne incognita* (Kofoid and White) Chitwood. *Plant and Soil, 303*, 207-216.

Tacconi, R., & Regazzi, D. (1990). Costi di coltivazione di colture intercalary resistenti a *H. schachtii. Informatore Fitopatologico* 40, 47-48.

Takehara, T. (2004). Principles and effects of soil disinfestation with hot water. *PSJ Soilborne Disease Workshop Report 22*, 22-37.

Tenuta, M., & Lazarovits, G. (2002). Ammonia and nitrous acid from nitrogenous amendments kill the microsclerotia of *Verticillium dahliae*. *Phytopathology 92*, 255-264.

Terefe, M., Tefera, T., & Sakhuja, P. K. (2009). Effect of a formulation of *Bacillus firmus* on root-knot nematode *Meloidogyne incognita* infestation and the growth of tomato plants in the greenhouse and nursery. *Journal of Invertebrate Pathology 100*, 94-99.

Ternisien, E. (1989). Etude des rotations culturales en bananeraie. II: Impact des cultures de rotation sur la production bananière et l'état sanitaire du sol. *Fruits 44*, 445-454

Ternisien, E., & Melin, P. (1989). Etude des rotations culturales en bananeraie. I. Bilan des cultures de rotation. *Fruits 44,* 373-383

Thames, W. H., & Heald, C. M. (1974). Chemical and cultural control of *Rotylenchulus reniformis* on cotton. *Plant Disease Reporter 58*, 337-341.

Thetford, M., & Kinloch, R. A. (1999). Galling response of herbaceous and woody landscape species grown in *Meloidogyne incognita*- and *M. arenaria*-infested soils. *West Florida Research and Education Center Research Report 1999-08.*

Thies, D. A., & Fery, R. L. (2002). Host plant resistance as an alternative to methyl bromide for managing *Meloidogyne incognita* in pepper. *Journal of Nematology 34*, 374-377.

Thurston, H. D. (2004). Ten thousand years of experience with sustainable plant disease control. *Plant Disease 88*, 550-551.

Tian, H. L., Riggs, R. D., & Crippen, D. L. (2000). Control of soybean cyst nematode by chitinolytic bacteria with chitin substrate. *Journal of Nematology 32*, 370-376.

Tiilikkala, K. (1991). Effect of crop rotation on *Globodera rostochiensis* and on potato yield. *EPPO Bulletin 21*, 41-47.

Timper, P., Gates, R. N., & Bouton, J. H. (2005). Response of *Pratylenchus* spp. in tall fescue infected with different strains of the fungal endophyte *Neotyphodium coenophialum*. *Nematology 7,* 105-110.

Todd, T. C. (1993). Soybean planting date and maturity effects on *Heterodera glycines* and *Macrophomina phaseolina* in Southeastern Kansas. *Journal of Nematology 25*, 731-737.

Toida, Y., Keereewan, S., & Puttirut, N. (1991). Assessment and prevention of mung bean damage caused by the root-knot nematode, *Meloidogyne javanica* in Thailand. *Japanese Journal of Nematology 21*, 6-10.

Trudgill, D. L., Phillips, M. S., & Alphey, T. J. W. (1987). Integrated control of potato cyst nematode. *Outlook on Agriculture 16*, 167-1 72.

Tsao, P. H., & Oster, J. J. (1981). Relation of ammonia and nitrous acid to suppression of *Phytophthora* in soils amended with nitrogenous organic substances. *Phytopathology 71*, 53-59.

Tzortzakakis, E. A., Adam, M. A. M., Blok, V. C., Paraskevopoulos, C., & Bourtzis, K. (2005). Occurrence of resistance-breaking populations of root-knot nematodes on tomato in Greece. *European Journal of Plant Pathology* 113, 101-105.

Uhlenbroek, J. H., & Bijloo, J. D. (1958). Investigations on nematicides. I. Isolation and structure of a nematicidal principle occurring in Tagetes roots. *Recueil des Travaux Chimiques des Pays-Bas* 77, 1004-1008.

Usha, K., & Patra, D. D. (2003). Medicinal and aromatic plant materials as nitrification inhibitors for augmenting yield and nitrogen uptake of Japanese mint (Mentha arvensis L. var. piperascens). *Bioresource Technology* 86, 267-276.

van Assche, C., & Uyttebroeck, P. (1983). Possibilities of microwaves in soil disinfestation. *EPPO Bulletin, 13*, 491-497.

van Bruggen, A. H. C., & Termorshuizen, A. J. (2003). Integrated approaches to root disease management in organic farming systems. *Australasian Plant Pathology 32*, 141-156.

van Damme, V., Hoedekie, A., & Viaene, N. (2005). Long-term efficacy of *Pochonia chlamidosporia* for management of *Meloidogyne javanica* in glasshouse crops. *Nematology 7*, 727-736.

van Wambeke, E., Wijsmans, J., & d'Hertefelt, P. (1983). Possibilities in microwave application for growing substrate disinfestation. *Acta Horticulturae 152*, 209-217.

Viaene, N. M., & Abawi, G. S. (1998). Management of *Meloidogyne hapla* on lettuce in organic soil with sudangrass as a cover crop. *Plant Disease 82*, 945-952.

Viglierchio, D. R. (1987). Biological criteria in management considerations for *Meloidogyne chitwoodi*. *Nematologia Mediterranea 15*, 65-70.

Walker, J. T. (1971). Populations of *Pratylenchus penetrans* relative to decomposing nitrogenous soil amendments. *Journal of Nematology 3*, 43-49.

Walker, J. T., & Melin, J. B. (1996). *Mentha × piperita*, *Mentha spicata* and effects of their essential oils on *Meloidogyne* in soil. *Journal of Nematology 28,* 629-635.

Wang, K.- H., McSorley, R., & Gallaher, R. N. (2004). Effect of *Crotalaria juncea* amendment on squash infected with *Meloidogyne incognita*. *Journal of Nematology 36*, 290-296.

Wang, K.- H., Sipes, B. S., & Schmitt D. P. (2001). Suppression of *Rotylenchulus reniformis* by *Crotalaria juncea, Brassica napus,* and *Tagetes erecta*. *Nematropica 31*, 237-251.

Wang, K.- H., Sipes, B. S., & Schmitt, D. P. (2002a). *Crotalaria* as a cover crop for nematode management: a review. *Nematropica 32,* 35-57.

Wang, K.- H., Sipes, B. S., & Schmitt, D. P. (2002b). Management of *Rotylenchulus reniformis* in pineapple, *Ananas comosus*, by intercycle cover crops. *Journal of Nematology 34*, 106-114.

Wang, K.- H., Sipes, B. S., & Schmitt D. P. (2003). Intercropping cover crops with pineapple for the management *of Rotylenchulus reniformis*. *Journal of Nematology 35*, 39-47.

Warrior, P., Rehberger, L. A., Beach, M., Grau, P. A., Kirfman, G. W., & Conley, J. M. (1999). Commercial development and introduction of DiTera^{TM}, a new nematicide. *Pesticide Science 55*, 343-389.

Weaver, D. B., Rodríguez-Kábana, R., & Carden, E. L. (1998). Velvetbean and bahiagrass as rotation crops for management of *Meloidogyne* spp. and *Heterodera glycines* in soybean. *Journal of Nematology 30*, 563-568.

Weaver, D. B., Rodriguez-Kábana, R., Robertson, D. G., Akridge, R. L., & Carden, E. L. (1988). Effect of crop rotation on soybean in a field infested with *Meloidogyne arenaria* and *Heterodera glycines*. *Annals of Applied Nematology 2*, 106-109.

Weibelzahl-Fulton, E., Dickson, D. W., & Whitty, E. B. (1996). Suppression of *Meloidogyne incognita* and *M. javanica* by *Pasteuria penetrans* in field soil. *Journal of Nematology 28*, 43-49.

Weingartner, D. P., & Shumaker, J. R. (1983). Nematicide options for northeast Florida potato growers. *Proceedings of the Florida State Horticultural Society 96*, 122-127.

Westphal, A., & Becker, J. O. (2001). Soil suppressiveness to *Heterodera schachtii* under different cropping sequences. *Nematology 3*, 551-558.

Whitehead, A. G. (1965). Taxonomy, distribution, and host-parasite relationships of the genus *Meloidogyne* Goeldi (Nematoda: Heteroderidae). Ph.D. Thesis, University of London.

Whitehead, A. G. (1977). Control of potato cyst nematode, *Globodera rostochiensis* Rol, by picrilonic acid and potato trap crops. *Annals of Applied Biology 87*, 225-227.

Whitehead, A. G. (1995). Decline of potato cyst nematodes, *Globodera rostochiensis* and *G. pallida*, in spring barley microplots. *Plant Pathology 44*, 191-195.

Whitehead, A. G. (1997). *Plant Nematode Control*. Wallingford, UK: CABI Publications.

Widmer, T. L., & Abawi, G. S. (2000). Mechanism of suppression of *Meloidogyne hapla* and its damage by a green manure of sudangrass. *Plant Disease 84*, 562-568.

Widmer, T. L., & Abawi, G. S. (2002). Relationship between levels of cyanide in sudangrass hybrids incorporated into soil and suppression of *Meloidogyne hapla*. *Journal of Nematology 34*, 16-22.

Widmer, T. L., Mitkowski, N. A., & Abawi, G. S. (2002). Soil organic matter and management of plant parasitic nematodes. *Journal of Nematology 34*, 289-295.

Williamson, V. M. (1998). Root-knot resistance genes in tomato and their potential for future use. *Annual Review of Phytopathology 36*, 277-293.

Wilson, G. F., & Caveness, F. E. (1980). Effect on rotation crops on survival of root-knot, root-lesion and spiral nematodes. *Nematropica 10*, 56-61.

Windham, G. L., Windham, M. T., & Williams, W. P. (1989). Effects of *Trichoderma* spp. on maize growth and *Meloidogyne arenaria* reproduction. *Plant Disease 73*, 493-494.

Xiao, J., Zhu, J., Chen, S., Ruan, W., & Miller, C. (2007). A novel use of anaerobically digested liquid swine manure to potentially control soybean cyst nematode. *Journal of Environmental Science and Health 42*, 749-757.

Zasada, I. A., & Ferris, H. (2004). Nematode suppression with brassicaceous amendments: application based upon glucosinolate profiles. *Soil Biology and Biochemistry 36*, 1017-1024.

Zasada, I. A., Meyer, S. L., Halbrendt, J. H., & Rice, C. (2005). Activity of hydroxamic acids from *Secale cereale* against the plant-parasitic nematodes *Meloidogyne incognita* and *Xiphinema americanum*. *Phytopathology 95*, 1116-1121.

Zasada, I. A., Rice, C. P., & Meyer, S. L. F. (2007). Improving the use of rye (*Secale cereale*) for nematode management: potential to select cultivars based on *Meloidogyne incognita* host status and benzoxazinoid content. *Nematology 9*, 53-60.

In: Organic Farming and Peanut Crops
Editors: D. C. Grossman, T. L. Barrios, pp. 65-102
ISBN: 978-1-60876-187-6

Chapter 2

Influence of Light Conditions on Biology and Chemistry in the Peanut Plant: Flavonoids and Spermidines from Peanut (*Arachis hypogaea*) Flowers and Studies of the Photoisomerization of Spermidine Conjugates

Victor S. Sobolev,[1] James B. Gloer,[2] and Arlene A. Sy[2]
[1]National Peanut Research Laboratory, Agricultural Research Service, U.S. Department of Agriculture, Dawson, Georgia 39842.
[2]Department of Chemistry, University of Iowa, Iowa City, Iowa 52242

Abstract

Early history and significance of the peanut crop is discussed. Annual world production of peanuts at 30 million tons makes this crop one of the most important agricultural commodities. Unusual physiology, inflorescence, and infructescence of the peanut plant make it an attractive object for scientific research. Many physiological processes in the peanut plant, such as vegetative growth, nitrogen fixation, and fertilization are associated with light. Present research revealed high light-sensitivity of the spermidines from the peanut reproductive organ, keel. A new spermidine triamide derivative has been isolated from peanut flowers and identified as N^1-acetyl-N^5, N^{10}-di-*p*-(*EE*)-coumaroylspermidine on the basis of detailed analysis of NMR, MS, and UV data. Two other spermidine conjugates, N^1, N^5, N^{10}-tri-*p*-(*EEE*)-coumaroylspermidine and di-*p*-(*EE*)-coumaroylspermidine, as well as four flavonoid conjugates (quercetin-3-glucoside, quercetin-3-glucuronide, isorhamnetin-3-glucoside, and isorhamnetin-3-glucuronide) that have been previously reported in organs of other plants, have been found in this study in peanut (*Arachis hypogaea* L.), a representative of the Leguminosae family, for the first time. The dynamics of photoisomerization in the spermidine conjugates have been investigated. These flavonoid and spermidine conjugates may play protective roles against pests. New spermidine conjugate may be involved in regulating the plant reproduction process.

Keywords: peanuts, *Arachis hypogaea*, groundnuts, *p*-coumaric acid, coumaroylspermidine, dicoumaroylacetylspermidine, tricoumaroylspermidine, quercetin glucoside, quercetin glucuronide, isorhamnetin glucoside, isorhamnetin glucuronide, photochemical isomerization, structure elucidation, NMR, HPLC-MS

INTRODUCTION

Origin and Use of Peanuts

Peanut or groundnut is a member of the genus *Arachis* of family Leguminosae. Cultivated peanut is classified based on the presence or absence of flowers in the main axis into two subspecies, A. hypogaea and A. fastigiata [1]. The only species in the genus of significant economic importance is *Arachis hypogaea* L., of which the seed is the most important plant part. *Arachis hypogaea* is an annual herb and is one of the few plant species that form underground fruits. The archaeological records support peanut cultivation between 300 and 2500 BC in Peruvian desert oases [2]. The cultivated peanut was likely first domesticated in the valleys of the Paraguay and Parana rivers in the Chaco region of South America. The first written account of the crop is found with the Spanish entry into Hispaniola in 1502, where the Arawak cultivated it under the name of mani [2]. Records from Brazil around 1550 show that the crop was known there with the name mandubi. As man became a foraging and then gathering species, the wild *Arachis* most surely contributed to his survival. In many of its native environments, *Arachis* contributes significantly to the perpetuation of the local ecology; that is, *Arachis* produces nitrogen for associated species, stability for soil particles, forage for animals, and seeds for various bird species [3]. Today, peanut is widely distributed and has become one of the world's most popular and universal crops, cultivated in more than 100 countries in all six continents. World peanut production totals approximately 30 million metric tons per year, with the U.S. being the world's third largest producer, after China and India [4]. A substantial proportion of total production is consumed by growers, without ever being recorded [2]. Peanuts, peanut oil, and peanut protein meals constitute an important segment of world trade in oilseeds and products. Peanut is the fifth most important oilseed in the world; a large percentage of peanut seed is utilized for the production of edible oil, whereas in the U.S. about 60% of production is processed in a variety of ways and eaten as food. At the beginning of the 20th century, American agricultural chemist George Washington Carver discovered over three hundred uses for peanuts [5], which had a significant impact on agriculture and created new markets for farmers. The stems and leaves of the peanut plant also are used as fodder for hogs and cattle in some areas, being economically more significant than the seeds at times of shortage of fodder for cattle in Asia and Africa. Dry peanut hulls (shells) are used as a degradable machine oil adsorbent and as a cat litter [6]. Potential uses of peanut seed skins (testa) have also been investigated [7, 8]. It has been shown that up to 35% of the oil in the skins can be recovered. In some cases, the oil can serve as a source of behenic and lignoceric long-chain fatty acids, which are used in body-building formulations and as ingredients in shampoos. After removal of the oil, the skins can be useful for making brandy, liqueur, and tea. Peanut skin oil extraction followed by tannin extraction also produces a protein-enriched product that could find application in mixed feeds for cattle consumption at higher concentrations relative to existing practice. A simple technique was also developed to use the skins in finishing decorative panels [7]. In 1947, Masquelier, a student at the University of Bordeaux, undertook a series of experiments and determined that peanut skins were not toxic. In the course of this research, he isolated a colorless substance from peanut skins, an oligomeric proanthocyanidin (OPC), and determined that it possessed vasoprotective potential [9]. In addition, peanut skin water-soluble extract contains other highly bioavailable procyanidins, as well as catechins, taxifolins, and an array of phenolic acids. OPC and similar compounds can be used for preventing and fighting the harmful biological effects of free radicals in the organism [8]. Peanut seeds and roots also contain resveratrol [10], a peanut phytoalexin precursor [11-13]. The importance of phytoalexins for the peanut plant survival is discussed in the next section of the chapter.

Role of Protective Phytoalexins in Peanut Physiology

Phytoalexins are low-molecular weight, biologically active metabolites produced by plants as defensive reactions to various exogenous stimuli, particularly fungal invasion [14]. Peanuts are host to approximately 50 genera of fungi [15], among which *Aspergillus flavus* and *A. parasiticus* are of particular agricultural significance due to their ability to invade peanut seeds and produce the carcinogenic aflatoxins [16]. Peanut can resist infection by pathogens, including *A. flavus* and *A. parasiticus* by synthesizing antibiotic stilbene-derived phytoalexins [12, 17-27] (Figure 1). Figure 2 shows sliced non-viable and viable peanut seeds, respectively, after 96 h of incubation with *A. caelatus*. Non-viable peanuts were completely colonized with the fungus, while only a few colonies were counted on viable peanuts. Apparently, under experimental conditions, viable peanuts were able to suppress fungal growth compared to non-viable peanuts. Sufficient water content in seeds is essential for effective production of the phytoalexins [24, 25]. However, the mechanism of peanut resistance to exogenous invasion has not been explored. Such natural phytoalexin-based mechanism of peanut resistance attracts the attention of researchers because, once understood, this mechanism of resistance could be manipulated to obtain a greater peanut resistance to pests. Such knowledge may be crucial for breeding new fungi-resistant peanut cultivars.

A number of stilbene phytoalexins from peanuts have been reported [12, 18-20, 23, 26, 27]. Figure 1 shows stilbenoids that have been isolated from peanut seeds. Resveratrol (**1**) and four prenylated stilbenoids **2-5** (Figure 1) were the first phytoalexins isolated from challenged peanut seeds in 1970-1980s [18-20, 26]. Under experimental conditions, these compounds represented the major elicited phytoalexins. However, one of the major pigmented compounds bearing unusual prenylated benzenoid and but-2-enolide moieties (**6**) (Figure 1), was overlooked and identified as a major component in infected seeds only in 2006 [12]. Recent isolation of six new stilbenoids **7-12** (Figure 1), some of which were detected in high concentrations [27], as well as preliminary data on the presence of new dimeric stilbenoids, has demonstrated that our knowledge about peanut resistance mechanisms is still limited and that systematic research on the issue is far from completion. Figure 3 shows typical HPLC results for extracts of peanut seeds inoculated with fungal spores. After 24 h of incubation, peanut seeds produced an array of stilbenoids at significant concentrations. It has also been demonstrated that other organs of the peanut plant are capable of producing the same phytoalexins [22]. However, a different group of prenylated stilbenes that are not found in seeds have been isolated from peanut root mucilage [23]. The principal low-polarity prenylated stilbene, mucilagin A, was suggested to play a role in regulating root-soil pathogen interactions [23]. It should be noted that all stilbenoids were isolated from the peanut plant strictly in the *trans*-configuration [12, 17-23, 25-27]. Several peanut stilbenoids display antifungal activity [18, 21, 26, 28].

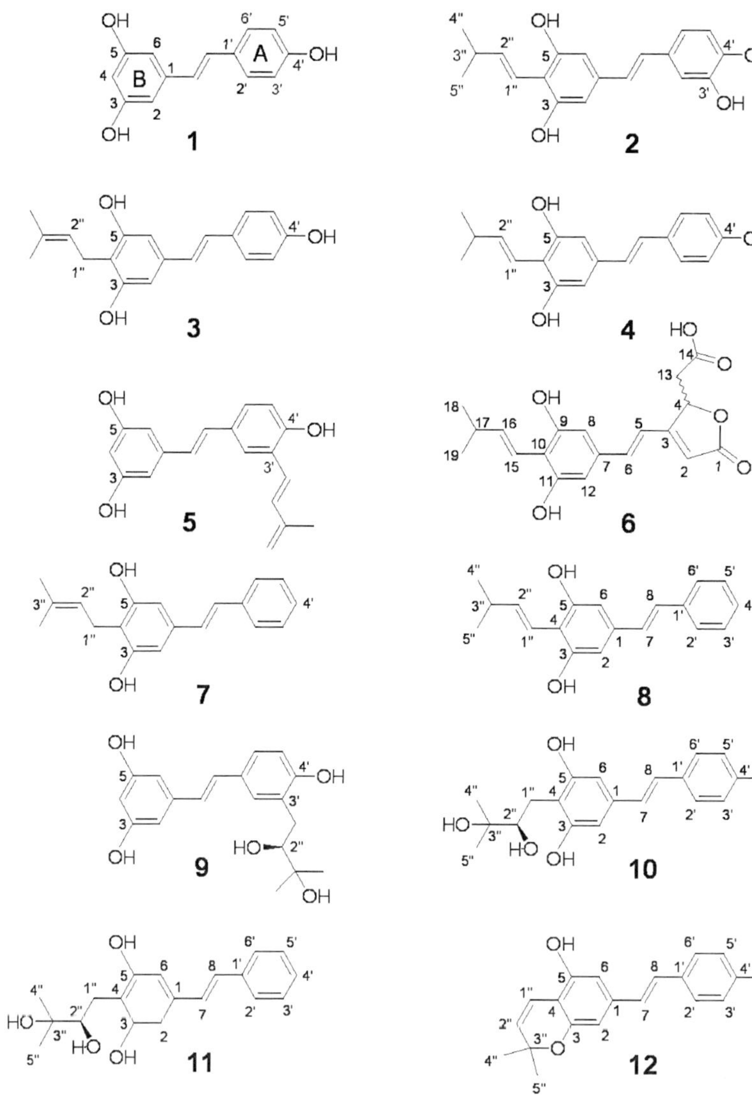

Figure 1. Structures of stilbenoids from peanut seeds: 1, *trans*-resveratrol; 2, *trans*-arachidin-1; 3, *trans*-arachidin-2; 4, *trans*-arachidin-3; 5, *trans*-3′-isopentadienyl-3,5,4′-trihydroxystilbene; 6, *trans*-SB-1; 7, chiricanine A (*trans*-4′-deoxyarachidin-2); 8, arahypin-1 (*trans*-4′-deoxyarachidin-3); 9, arahypin-2 [*trans*-3′-(2″,3″-dihydroxy-3″-methylbutyl)resveratrol]; 10, arahypin-3 [*trans*-4-(2″,3″-dihydroxy-3″-methylbutyl)resveratrol]; 11, arahypin-4 [*trans*-4-(2″,3″-dihydroxy-3″-methylbutyl)-4′-deoxyresveratrol]; 12, arahypin-5.

Figure 2. Sliced peanut seeds after 96 h of incubation with *A. caelatus* NRRL 25528 at 30 °C; **A**, non-viable seeds; **B**, viable seeds.

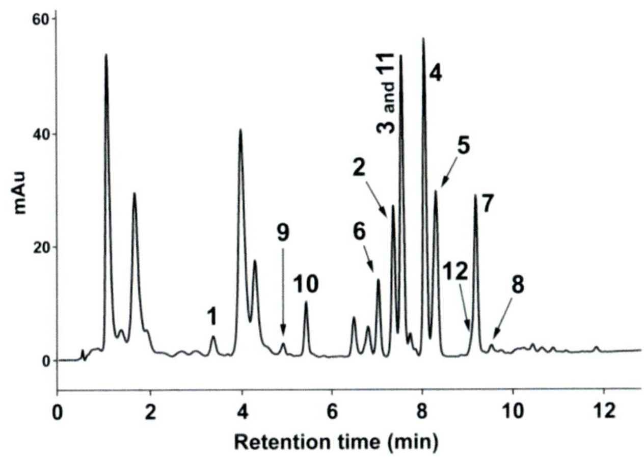

Figure 3. HPLC results for peanut seed extracts: A, non-viable uninoculated seeds after 24 h of incubation; B, viable uninoculated seeds after 24 h of incubation; C, viable seeds after 24 h of incubation with *A. caelatus*; D, viable seeds after 96 h of incubation with *A. caelatus*; 1, resveratrol; 2, arachidin-1; 3, arachidin-2; 4, arachidin-3; 5, *trans*-3′-isopentadienyl-3,5,4′-trihydroxystilbene; 6, SB-1; 7, chiricanine A; 8, arahypin-1; 9, arahypin-2; 10, arahypin-3; 11, arahypin-4; 12, arahypin-5. Stilbenoids 1-12 were all detected in the *trans*-configuration. The chromatograms were recorded at 315 nm.

Pathogens attack all parts of the peanut plant throughout the growing season, restricting normal peanut growth and development [29]. Most of the pathogens that attack peanut are of fungal origin [30]. Sobolev *et al.* investigated peanut seeds for their ability to produce phytoalexins in response to infection by various *Aspergillus* strains. The experimental design allowed for the study of phytoalexin production in peanut seeds layer-by-layer. Thin layers were dissected from different depths of the kernel starting from the infected area down to healthier tissues. Six peanut phytoalexins, *trans*-resveratrol (**1**), *trans*-arachidin-1 (**2**), *trans*-arachidin-2 (**3**), *trans*-arachidin-3 (**4**), *trans*-3'-isopentadienyl-3,5,4'-trihydroxystilbene (**5**), and SB-1 (**6**) (Figure 1) were detected. It was demonstrated that different fungal strains from the genus *Aspergillus* were able to elicit intensive phytoalexin synthesis in a localized area close to the infected wound of a peanut kernel under favorable conditions to both the host and the fungus. Higher phytoalexin concentrations were accumulated with longer incubation times, and the composition of phytoalexins varied significantly by layer. Resveratrol (**1**) (Figure 1) is most likely a building block for other stilbenoids that seem to emerge after its appearance. There was a significant difference in phytoalexin production elicited by some fungal isolates. No association was observed between phytoalexin production and toxigenic potential of fungal strains that elicited the production in mature peanut kernels.

Dorner *et al.* [25] demonstrated that, as peanut seed water content decreased during drought and temperature stresses, the capacity of seeds to produce phytoalexins declined, resulting in increased levels of fungal infection and aflatoxin contamination. Late-season drought can lead to dangerous aflatoxin contamination [25, 30, 31] and insect injury, especially by the lesser cornstalk borer (LCB), which can exacerbate aflatoxin contamination in peanuts [31, 32]. Invasion by *A. flavus* and *A. parasiticus* and aflatoxin production in peanuts subjected to water-stress first occurs in small, immature peanuts [25], suggesting that mature peanut pods possess more effective resistance mechanisms that inhibit fungal growth [33]. For many diseases, highly effective resistance is not known, so quantitatively expressed resistance is of significant interest.

There are over 100 insect pest species that can inflict high-impact on peanuts [30]. Aflatoxin concentrations in seeds from pods injured by insects can be dramatically higher than in seeds from uninjured pods. Insect injury to peanut pods may result in aflatoxin contamination in seeds under conditions that normally do not favor fungal infection or aflatoxin production in uninjured pods.

Throughout the 20th century, scientists have exploited natural resistance to improve crop varieties. As a result, breeding for plant resistance to plant pathogens and insects has been one of the major achievements in the control of several important diseases in peanuts. Host plant resistance is an efficient, economical, and environment-friendly approach used to manage many pests and diseases of agricultural crops. In spite of the large number of cultivars available to growers, the peanut crop has been characterized as being genetically vulnerable to diseases and insect pests [34, 35].

Efficient selection for resistance during the breeding process has been facilitated by linkages to markers [36]. Chemical markers can be useful in efforts to combine genes with similar resistance phenotypes. It has been demonstrated that peanut phytoalexins can serve as chemical markers for runner type cultivars [37]. Runner cultivars have become the dominant peanut type grown after the introduction in the 1970's of a new cultivar, Florunner, which was responsible for a dramatic increase in peanut yields [30]. The current dominant cultivar is Georgia Green, which became popular because of its field resistance to tomato spotted wilt virus (TSWV) [30]. At present, runner type cultivars account for 80% of total production in the USA [30]. Resistance of peanuts to TSWV is one of the major issues in the US peanut industry. It was shown that peanuts grown under real field conditions were capable of producing significant concentrations of phytoalexins [37]. Phytoalexin production was highest in insect-damaged peanut pods. A correlation has been noted between total phytoalexin production and published genotype resistance to TSWV and LLS diseases. Damaged pods of the peanut genotypes most resistant to TSWV and LLS consistently produced higher concentrations of phytoalexins compared to susceptible genotypes. There was no significant difference in phytoalexin production by undamaged peanut pods of all tested genotypes. *Trans*-arachidin-3 (**4**) and *trans*-resveratrol (**1**) (Figure 1) were the major phytoalexins produced by insect-damaged pods of all tested peanut genotypes. Therefore, these compounds seem to be the stilbenoids of choice when investigating peanut genotypes for their ability to synthesize phytoalexins, and possibly for overall peanut resistance to diseases. Stilbene phytoalexins may be considered potential chemical markers in breeding programs for disease-resistant peanuts. The presence of stilbenoids in organs of the peanut plant suggests that they may play protective roles for the entire plant. There is no evidence that stilbene phytoalexins are directly involved in peanut resistance to viral diseases such as TSWV. However, the ability to synthesize high levels of phytoalexins might be an indication of genetically inherited vigor resulting in superior growth and yield. Such peanuts may have greater ability to maintain multiple physiological resistance functions under challenged conditions. Understanding the processes triggering resistance will be useful for manipulating resistance genes for greater effectiveness and stability of resistance.

It should be also noted that peanut phytoalexins may also have potential as natural antioxidants with beneficial human and animal health properties. One of the major peanut stilbenoids, resveratrol, is known to possesses diverse beneficial health properties, including anticancer properties [38-41].

Roles of other important factors in aspects of peanut physiology are discussed in the following sections.

Role of Light in Vegetative Peanut Growth

The absorption of solar radiation by the peanut plant, in addition to being the energy source for photosynthesis, is a major factor in regulating plant growth and development. Irradiance (light intensity), quality of light, and photoperiod influence the way the plant grows. A unique characteristic of the peanut plant is the nyctinastic movement (sleep movement) of the leaflets. The leaf blade consists of four oval to oblate leaflets attached to the midrib by small articulations which allow for movement. During dark periods and hot sunny days, the paired leaflets are close together in a vertical position, and on a normal day, leaflets are separated from each other in a horizontal position. Blue light or far red irradiation cause the leaflets to open [42, 43]. It has been shown that in darkness, or as solar radiation was decreased, the cotyledonary branches of runner type plants grew more erect and returned to horizontal growth in light. Also, with reduced irradiance or blue light, the main axes and lateral branches were longer than at high irradiance or in blue plus far red irradiation. As less solar radiation was received and the treatment prolonged, plants produced less shoot dry weight and fewer mature pods and seed [44]. Shaded plants also had a slower rate of vegetative and reproductive growth. The shaded plants apparently compensated for the treatment by producing fewer seeds, but with the largest mean seed weight. This research agrees with the result that shading in the post-flowering period caused a reduction in the percentage of extra large kernels, but no significant difference in yield of individual shading treatments occurred [45]. Many leguminous species are grown under relatively high plant population densities > 250 000 plants ha^{-1}. Some peanut cultivars were successfully grown even at higher densities reaching 352 000 plants ha^{-1} [46, 47]. It has been suggested that these plants can tolerate high population densities because their nyctinastic leaf movements allow their aerial parts to receive directly the last light of the day, which increases the P_{fr} level in the plants before the onset of darkness, causing them to develop like unshaded plants. Measurements of irradiance above and below the canopy of a densely planted crop of peanut (*Arachis hypogaea fastigiata* Waldron) showed that, during daytime, when the leaf laminae were horizontally oriented, only 10% of the incident light reached the ground, while after sunset, with the leaves in the vertical position, 70 to 90 % of the incident light could penetrate to the ground. Natural light, available to the plants after this time, was capable of inducing the straightening of hypocotyl hooks of dark-germinated seedlings, a typical red light requiring photomorphogenic reaction. Laboratory experiments showed that peanut plants, receiving a daily exposure of 5 min to red light immediately after the end of the photoperiod developed significantly more branches and more reproductive organs than plants receiving 10 min of exposure to far red light. The greater adaptability of the peanut plant to high population density was demonstrated by comparing the above peanut species performance with that of maize (*Zea mays* L.), okra (*Abelmoschus esculentus* Moench), safflower (*Carthamus tinctorius* L.), and sesame (*Sesamum indicum* L.) under a high and a low population density. Although crowding caused the plants of all five species to grow taller, with fewer branches and fewer reproductive organs, the harvest index of the peanut species remained unaffected by the population density, while that of the remaining four species was significantly reduced by the higher population density [48]. Among the climatic factors tested by Gautreau [49], total radiation received by the plants was a predominant factor in determining growth and fruiting. The way that light affects plant growth habits suggests that it may also alter endogenous levels of gibberellic acid-like phytohormones [43, 50]. In relation to flowering and fertilization, Chhabra and Malik [51] showed that red light hastened pollen tube growth and far red light reversed this effect. Blue light inhibited pollen tube growth.

Role of Light in Nitrogen Fixation

Peanuts and other legumes can form symbiotic associations with soil bacteria of the genus *Rhizobium*. The bacteria infect the plant roots, proliferate, and invade the root cells [52]. The root cells then proliferate to form the nodules where molecular nitrogen from the atmosphere is fixed. These nodules contain bacterial cells (bacteroids), host plant cells, and leghemoglobin. The leguminous plant by itself cannot fix nitrogen, but bacteroids and some free-living rhizobia can fix nitrogen under proper environmental conditions [53, 54]. Symbiosis between plant and bacteria does not guarantee vigorous nitrogen fixation. In fact, bacteria can sometimes infect the roots and form nodules, but ineffectively fix nitrogen [55-59]. Effective nitrogen fixation depends on specific elements of the relationship between host plant and bacteria. One of the major known contributions of the shoot to nitrogen fixation is photosynthate supplied to the nodule [60-62], and one of the environmental factors limiting photosynthesis and plant growth and development is light. Diurnal effects on nitrogen-fixing activities have been found with maxima occurring near the time of maximum light intensity [61, 62]. Peanuts in the field showed two maxima: one near the end of the night period, and another toward the end of the day [63]. Few or no nodules develop on seedlings in the dark [64]. Shading decreased nitrogen fixation, while supplemental light increased fixation, and nodule-specific nitrogen-fixing activity was similarly affected [61, 62]. Increasing the level of carbon dioxide in the peanut canopy increased nitrogen fixation by increasing nodule number and mass. This was attributed to increased photosynthesis and available photosynthate [61]. Plants grown at low and high light intensities and then transferred to the opposite treatment showed an increase and decrease in nitrogenase activity, respectively [65]. Photoperiod also affects legume nodulation, with long days promoting more and larger nodules than short days

[65, 66]. This result suggests a direct photosynthetic effect on symbiosis since this occurs regardless of the photoperiodic requirements of the host for flowering [66]. However, as discussed above, peanut plants produce more fruit under comparatively short days. There is also a nonphotosynthetic effect of light quality on nodulation. Exposure of the shoot or root of some legumes to far red light inhibits nodulation, while red light reverses this effect, which suggests the involvement of the phytochrome system [65, 66].

Role of Light and Other Factors in Flowering

The peanut plant is unusual because it flowers above ground and pods containing one to five seeds are produced underground. As a legume, *Arachis hypogaea* is unusual in its genus being polyploid (4x=40). It can interbreed only with another species *A. monticola*, the probable wild progenitor of the crop [2]. The cultivated peanut plant is erect or prostrate, usually 15 to 60 cm tall. It has a well developed tap root system with many lateral roots. Peanuts are indeterminate in growth habit, which means that vegetative and reproductive peanut growth occur simultaneously. The plant must produce enough nutrition to continue vegetative growth, as well as provide nutrition for seed development [67].

Peanut is a self-pollinating plant with unusual inflorescence. The flower contains five petals: a standard, two wings, and two petals fused to form a keel (Figure 4). The flower has 10 stamens, two of which are usually not fully developed; the pistil consists of an ovary, style, and stigma. Important phonological development stages of the peanut plant are described in detail by Munger *et al.* [68]. The first flowers appear 4 to 6 weeks after emergence. Flowering continues for 6 or more weeks. Any environmental stress may interrupt flowering. Often, excessively high temperatures and drought stress can result in abortion of flowers and sterilization of pollen. These factors greatly influence yield potential. There may be several flowers in an inflorescence on the plant, but only one opens on a given morning, and there is an interval from one to several days between the openings of successive flowers [69]. Athesis and pollen shedding occur at sunrise with pollination taking place within the closed keel of the flower. Fertilization occurs 8 to 9 hours after pollination [70]. If fertilization of the ovule is successful, the flower fades and the ovary elongates to become the peg which begins to grow downward, drives into the soil where it matures into a pod with seeds. Under favorable temperatures, nutrient, and moisture conditions, mature fruit develops in 9 to 10 weeks. The mature pod is an indehiscent legume containing one to five seeds, enclosed in papery testas. The seeds do not contain an endosperm, but have two large cotyledons, an epicotyl with three meristems, a hypocotyl, and a primary root. Diseases and damage by insects can effectively reduce leaf or root surface area, or flower integrity, which also may result in slower maturation and lower yields.

In addition to self-fertilization, the peanut flower may be cross-pollinated by bees [71-73]. In general, bee-pollinated flowers are typically zygomorphic; mechanically strong, with adequate facilities for landing and a surface that gives a good foothold; intricate and semi-closed, especially those visited by larger bees. They tend to have brightly colored corolla, generally yellow or blue; and nectar guides. The nectar is hidden, with nectaries characteristically situated at the base of the corolla tube. These flowers typically have a fresh, but not strong odor. In general, bee pollinated flowers have a distinctive pattern by which bees can recognize them. Because bees' vision covers the range from 300 - 650 nm (in contrast to humans - 400-700 nm) flowers often have distinctive patterns that are visible only in UV light. From the above description of bee-pollinated flowers, the value of peanut flowers to bees is not clear. Apparently, there is no functional nectary within the flower. Peanut flowers, when irradiated with long-wave UV light, do not produce fluorescence visible to human eyes. The flower is not mechanically strong and does not have characteristic patterns of typical bee-pollinated flowers. It has been observed [74] that varietal differences exist for cross-pollination in relation to the protrusion of the stigma out of the keel. It was reported [75] that two cultivars have a structure in the flower that obstructs self-pollination, but which facilitates cross-pollination by bees. Leuck and Hammons [76] obtained no hybrids from caged plants, but got a significant number from plants not caged; at least 80 percent of the peanut flowers in the open were actually tripped for pollen by bees [77]. Girardeau and Leuck [78] showed that caged flowers not manipulated in any way produced 4 to 11 percent fewer fruits than hand-flexed or water-drip manipulated flowers.

Peanut flowers have been considered resistant to pests [79, 80] with one exception being *Heliothis* spp. caterpillars [81]. Compounds that may be responsible for such resistance have not been investigated.

Role of Flavonoids and Spermidines in Plants; Association of Flavonoid and Spermidine Formation and Action with Light Exposure

Flavonoids are pigments that are biosynthesized from phenylalanine and are crucial for plants [82, 83]. Over 4000 chemically unique flavonoids have been identified in plant sources [82]. Flavonoids generally display vivid colors associated with flower petals, and most emit intense fluorescence when they are exposed to UV light. Flavonoids are ubiquitous to plant cells that contain chlorophyll, and are found in different organs of plants in free or bound forms [82, 83]. Their formation normally depends on light so that they are mainly concentrated in the outer tissues, and can act as light screens [82]. One of their major functions in plants is protection, as they kill or inhibit growth or reproduction of several pathogenic bacterial, fungal, and viral strains, as well as protozoans. Flavonoids may act as precursors of toxic compounds. However, the toxicity of flavonoids to herbivorous animals is low [82, 84]. Flavonoids regulate plant growth by inhibition of the exocytosis of the auxin indolyl acetic acid, as well as by induction of gene expression, and they influence other biological cells in numerous ways. Flavonoids readily participate in biological oxidation-reduction processes, and therefore are effective catalysts of electron transfer reactions. Flavonoid content is often used by scientists for taxonomic purposes [84]. It is important to note that, besides their crucial role in plants, flavonoids are major functional components of many herbal recipes. Flavonoids are used to treat some common diseases due to their proven ability to inhibit specific enzymes, to neutralize free radicals, and to simulate some hormones and neurotransmitters [84]. Flavonoids have also been reported to possess anti-inflammatory, vascular, oestrogenic, and antitumor activity [85].

Hydroxycinnamic acid amides (HCAA) are a widely distributed group of secondary metabolites that are concentrated in the floral parts of plants [69, 86-94]. Several HCAA have been isolated and identified [88-93, 95]. Some fundamental functions have been attributed to these amides, including protection against fungi, bacteria, viruses, and insects [96-104]. N^1-coumaroylspermidine is structurally similar to acylpolyamines found in spider and wasp venoms, which are known to block arthropod glutamate receptors. The ability of this coumaroylspermidine to attenuate synaptic transmission at insect neuromuscular synapses supports the suggestion that plant-derived phenolic polyamines may serve as natural insecticides [103]. HCAA also influence growth, cell multiplication, floral induction, flower formation, sexual differentiation, tuberization, cell division, and cytomorphogenesis [92, 96, 101, 105-108].

It has been found that HCAA accumulation in strawberry (*Fragaria ananassa* Duch.) is associated with plant irradiation with light [109]. Different associations of free amines and conjugates were observed during floral development as compared to floral induction. During the whole period of floral development in normal plants, phenylethylamine was observed to be the predominant amine. Conjugated spermidine was the predominant amide during the earliest stages of floral differentiation. Phenylethylamine conjugates were the predominant amides in the days preceding fertilization. The levels of these compounds decreased drastically after fertilization. In plants cultivated under long-days, free phenylethylamine, HCAA, and bound aromatic amines did not appear. Free polyamine and bound polyamine levels were low and only small changes were observed during the entire period of vegetative development. Sterile flowers of strawberry can be distinguished from fertile flowers by their lack of HCAA and a decrease in free phenylethylamine [109]. In *Xanthium strumarium,* a short-day plant, exposure of leaves to successive inductive nights resulted in a rise in conjugated spermines [110]. HCAA rose sharply after one inductive long night, remained high during the second cycle, and then declined rapidly during the third and fourth inductive cycles. Approximately eight days after the induction, the initial buds appeared, which also correlated well with the behavior expected of a floral stimulus. It was proposed that during the early events of flowering in *Xanthium strumarium*, a movement of polyamines occurred from young, expanding leaves to the buds and developing inflorescence. Such a mechanism would be consistent with the data obtained from studies on tobacco [111, 112], chrysanthemum [113], and strawberry [109]. If HCAA are causally involved in any type of morphogenesis, it may be possible in the near future to alter development by manipulating the endogenous levels of polyamines in transgenic plants using tissue-specific or inducible promoters [109].

Figure 4. Peanut flower. 1 – standard; 2 – one of two symmetrical wings (one of the wings is removed for a better observation of the keel); 3 – keel (2 fused petals that enclose pistil and stamens); 4 – hypanthium.

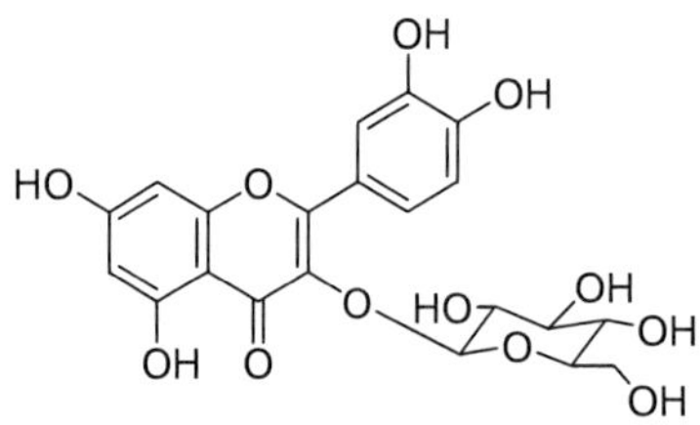

Quercetin-3-glucoside (**13**)

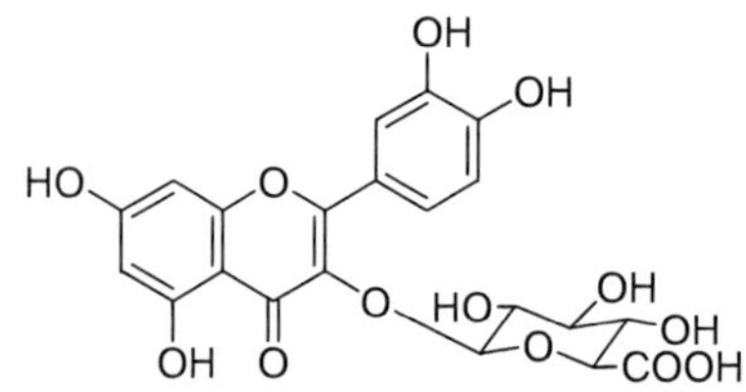

Quercetin-3-glucuronide (**14**)

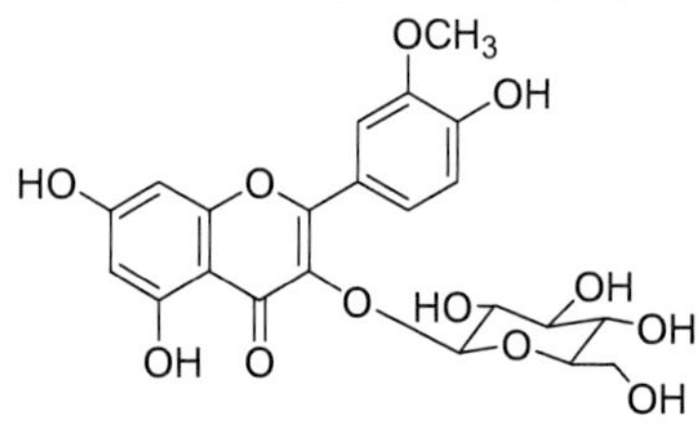

Isorhamnetin-3-glucoside (**15**)

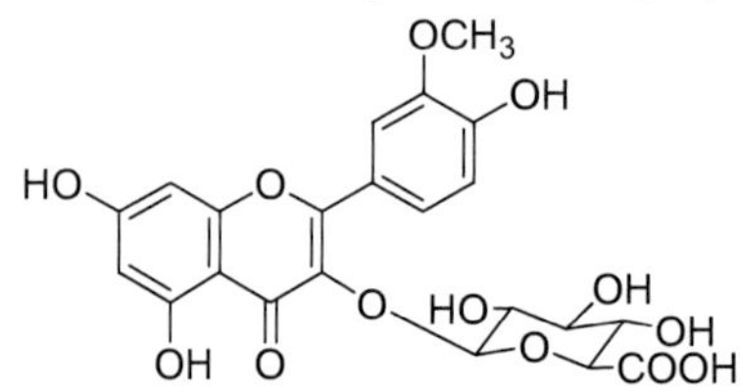

Isorhamnetin-3-glucuronide (**16**)

Figure 5. Structures of peanut flavonoid conjugates.

Hydroxycinnamic acid-spermidine conjugates have proven useful for chemotaxonomic classification in several plant families [88]. In addition to diverse functions in plants, three HCAAs, particularly N^{1}, N^{5}, N^{10}-tri-*p*-coumaroylspermidine (**19**) (**Figure 6**), have been reported to appreciably inhibit HIV-1 protease that is essential for the life-cycle of HIV [114]. The latter compound also demonstrated potent activity against *Helicobacter pylori*, a major etiological agent in gastroduodenal disorders [115]. It should be noted that about 60% of the world's population relies almost entirely on plants for medication. In this respect, any thorough investigation of abundant cultivated plants should be encouraged.

The following sections describe in detail results of our recent research on peanut flower constituents and their potential role(s) in the plant.

Rationale for Further Research on Peanut Flower Constituents

Peanut is an economically and nutritionally important crop. Sound seed yield depends heavily on high resistance to pests and uncompromised plant fertility. However, peanut flower constituents that may be responsible for the resistance and fertility have not been studied. The role of light in peanut flowering has been articulated mainly as the results of observations without investigation of specific photochemical mechanisms or compounds involved.

The goal of this research was to investigate the presence of major phytochemicals in different organs of the peanut flower that may be responsible for peanut fertility and protection against pests, and to explore any photochemical properties that might be associated with such metabolites.

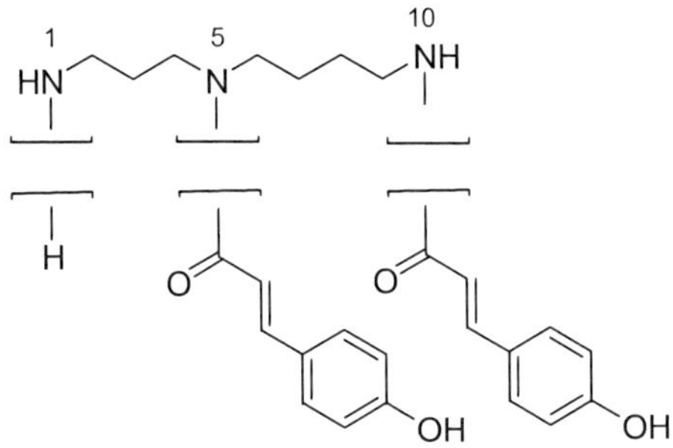

di-*p*-(*EE*)-coumaroylspermidine (**17**)

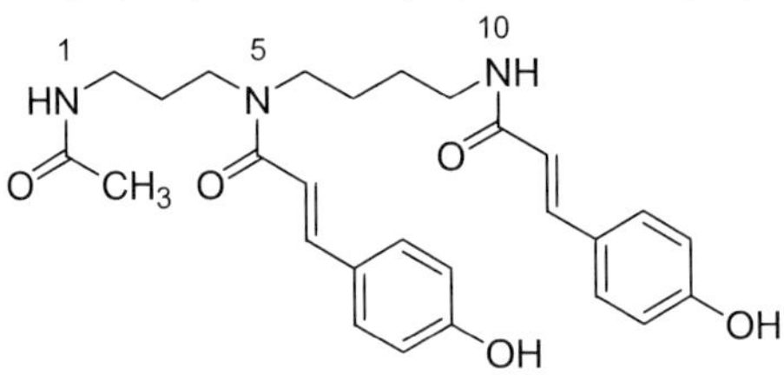

N[1]-acetyl-*N*[5], *N*[10]-di-*p*-(*EE*)-coumaroylspermidine (**18**)

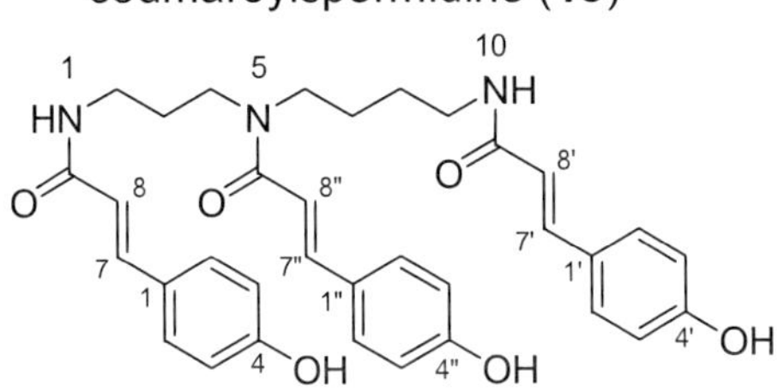

N[1], *N*[5], *N*[10]-tri-*p*-(*EEE*)-coumaroylspermidine (**19**)

Figure 6. Structures of peanut spermidine conjugates.

Experimental Procedures

Reagents, Materials and Basic Apparatus. HPLC-grade solvents used in the preparation of mobile phases, sample extraction, separations on silica gel, and isomerization were obtained from Fisher (Suwanee, GA). Silica gel for thin-layer chromatography (TLC; Silica Gel 60H) and for column chromatography (Silica Gel 60, 0.063-0.200 mm) were purchased from EM Science (Gibbstown, NJ). Na_2SO_4 was purchased from J.T. Baker. A high-speed blender with 1 L glass jar (13000 rpm; General Electric), a rotary evaporator (Rotavapor-R; Brinkmann; bath temperature, 40°C), and a model T-9 ultrasonic bath (L&R) were used in the research. A 2-L Kimax Büchner-type filtering funnel with 10-15 μm fritted disk was used as a column for preparative chromatography (i.d. 125 mm; Fisher).

Reference Compounds. *p*-Coumaric acid was purchased from Aldrich. ESI-MS/MS (MS^2) and UV data were obtained for the above commercial compound to serve as a comparison standard for identification of *p*-coumaric acid extracted from peanut flowers. ESIMS analysis of this phenolic acid gave an $[M - H]^-$ ion at *m/z* 163 and UV maxima at 293 sh, 300 sh, and 308 nm.

Plant Material and Processing. The following peanut breeding lines were used:

Runner market type: Tifrunner (resistant to Tomato spotted wilt virus), G02C (resistant to White mold; some resistance to *Cylindrocladium* Black Rot; nematode-, *M. arenaria*-, and Leaf spot- susceptible genotype; severe pod galling with *C. parasiticum*), C724-19-15 (resistant to *M. arenaria*; no severe pod galling with *C. parasiticum*; nematode-resistant genotype), AT29-1112, and AT3-1114. Virginia market type: VC-2, AT3081 R, Gregory, AT3085RO, experimental breeding lines termed in this publication A and B. Four plants in each pod with commercial pod soil were grown in a temperature-, illumination-, and humidity-controlled green house. The plants were treated in accordance with good agricultural practice. Peanut flowers, with about 10-mm long hypanthia, were collected every morning (from 7 to 9 am) from August 1, 2007 to August 21, 2007. Whole peanut flowers were frozen immediately after their collection to stop possible enzymatic reactions.

Extraction and Purification. For analytical purposes, 20 g of whole peanut flowers were extracted with 100 mL of MeOH in a high speed blender for 2 min. Filtered aliquots of the extract were used for direct determination of target constituents by HPLC.

For preparative isolation of conjugates, 800 g of peanut flowers combined from 11 breeding lines were extracted portion-wise (100 g each) with 2.4 L of MeOH (300 mL for each portion) in a high-speed blender for 2 min. The combined mixture was filtered through a glass-fiber filter in a Büchner-type funnel under reduced pressure. The solid residue was resuspended in 2.0 L of MeOH, and the extraction procedure was repeated twice as described above. The combined extract solutions were filtered through a filter paper and defatted twice with 0.5 L of *n*-hexane. The MeOH layer was evaporated nearly to dryness with 100 g of silica gel for column chromatography using a rotary evaporator. To remove excess water from the silica gel, an appropriate amount of acetonitrile was added close to the end of evaporation, and the mass was evaporated to dryness. The silica gel appeared as free-flowing powder. For rough isolation of target compounds, a 2-L filtering funnel packed with 330 g of silica gel for TLC (suspended in $CHCl_3$) was used as a column for preparative chromatography (column #1). Dry silica gel with the flower extract adsorbed was resuspended in $CHCl_3$ and applied to the column. The column was subsequently eluted with 1 L of $CHCl_3$, 2 L of EtOAc, 1.5 L of EtOAc-acctone (5:1, v/v), 2 L of EtOAc-acetone (1:1, v/v), 1.5 L of acetone, and 1.5 L of MeOH. Twenty-five fractions were collected from the column and analyzed by HPLC. Fractions containing target flavonoid conjugates (Figure 5) and spermidine conjugates (Figure 6) were separately combined, evaporated to dryness with a rotary evaporator, and subjected to further purification. Combined flavonoid fractions were dissolved in MeOH, filtered, and subjected to a final purification on preparative HPLC column (XTerra Prep RP_{18} OBD, 5 μm, 19×100 mm) using gradient mobile phase #1. Combined fractions containing flavonoid conjugates were evaporated with a rotary evaporator to a point where almost all of the MeOH was removed. The remaining aqueous mixture containing the target compounds was extracted 4 times with EtOAc (H_2O-EtOAc ratio 1:3, v/v). The combined EtOAc layers were evaporated nearly to dryness with a rotary evaporator. The residue was transferred into a 15-mL vial with MeOH and evaporated to dryness with in a stream of N_2. Combined fractions containing the spermidines were resuspended in $CHCl_3$, sonicated in an ultrasonic bath for 1 min, and applied to silica gel column #2 (40 x 350 mm). The column was subsequently eluted with $CHCl_3$ (300 mL), EtOAc (400 mL), EtOAc - acetone (300 mL + 75, mL), acetone (400 mL), $CHCl_3$ – MeOH (300 + 75 mL), and MeOH (400 mL). Fourteen fractions were collected. Tricoumaroylspermidine was mainly eluted with the EtOAc, EtOAc - acetone, and acetone eluting solvents. Dicoumaroylacetylspermidine was mainly eluted with the $CHCl_3$ - MeOH mixture. Fractions containing tricoumaroylspermidine and dicoumaroylacetylspermidine were redissolved in EtOAc - acetone (2:1, v/v) and separately applied to two Al_2O_3 columns (27 x 80 mm each; columns #3 and #4). The columns were subsequently eluted with EtOAc - acetone (50 + 50 mL), acetone (250 mL), MeOH (100 mL), MeOH - H_2O - HCOOH (196 + 3.5 + 0.5 mL). The spermidines were eluted from both columns with the last solvent mixture. Fractions containing spermidine conjugates were evaporated to dryness. The fraction containing tricoumaroylspermidine was redissolved in MeOH, filtered, and subjected to a final purification on preparative HPLC column (XTerra Prep RP_{18} OBD, 5 μm, 19×100 mm) using isocratic mobile phase #3. Fractions from the HPLC column containing tricoumaroylspermidine were evaporated with a rotary evaporator to a point where almost all CH_3CN was removed. The remaining aqueous mixture containing the target compound was extracted 4 times with EtOAc (H_2O-EtOAc ratio 1:4, v/v). The combined EtOAc layers were evaporated nearly to dryness with a rotary evaporator. The residue was transferred into a 4-mL vial with MeOH and evaporated to dryness with in a stream of N_2. Fractions from aluminum oxide column #4 containing dicoumaroylacetylspermidine were combined, evaporated with a rotary evaporator to a point where almost all MeOH was removed. The remaining aqueous mixture containing the target compound was extracted 4 times with EtOAc (H_2O-EtOAc ratio 1:3, v/v). The combined EtOAc layers were evaporated nearly to dryness with a rotary evaporator. The residue was transferred into a 15-mL vial with MeOH and evaporated to dryness with in a stream of N_2. For additional purification of the compound, the residue was dissolved in EtOAc - acetone (2:1, v/v) and applied to silica gel column #5 (30 x 120 mm). The column was subsequently eluted with $CHCl_3$-MeOH mixtures of increasing polarity (250 + 5, 250 + 10, 250 + 15, 250 + 25, 250 + 40, 250 + 50 mL). Fractions containing the target compound were combined, evaporated with a rotary evaporator to dryness, redissolved in MeOH and subjected to final purification on a preparative HPLC column (XTerra Prep RP_{18} OBD, 5 μm, 19×100 mm) using isocratic mobile phase #2. Isolation of the compound was performed as described above for tricoumaroylspermidine.

For extraction of peanut flower organs, the flowers were separated into four parts: standard, wings, keel, and hypanthium (Figure 4). Separated parts were ground in an agate mortar with an equal amount (w/w) of Celite 545. The mixtures were extracted with triple amount of MeOH (w/v) for 30 min with periodic (every 5 min) vortexing for 30 sec. Each extract was filtered through a glass-fiber filter. From 5 to 30 μL of each filtrate were analyzed by HPLC.

HPLC-DAD Analyses. Analyses were performed using an HPLC system equipped with a pump (model LC-10ATvp; Shimadzu), a DAD covering the 200-600 nm range (model SPD-M10Avp with Shimadzu Client/Server software, version 7.3), and an autosampler (model 717 plus; Waters). The separations of whole flower extracts and purified flavonoid conjugates were performed on an XTerra MS C18 analytical column (3.5-μm particle size, 4.6 × 150 mm; Waters). H_2O (A), MeOH (B), 2% of 2%-HCOOH in H_2O (C), and CH_3CN (D) were used in the following gradient: initial conditions, 68% A/ 30% B/ 2% C, increase linearly to 0% A/ 98% B/ 2% C in 22 min, hold isocratic for 5 min, decrease to initial conditions in 0.01 min. For separations of spermidine conjugates and their photoisomerization products, an isocratic mobile phase composed of 74% of (A), 2% of (C), and 24% of (D) was used. The flow rate in both cases was 1.0 mL/min. The column was maintained at 40°C in a column heater (model 105; Timberline Instruments). The eluate from the DAD was split with a T-unit (Upchurch Scientific; Oak Harbor, WA) for optimal MS performance. Flow rate through the ESI probe was set at 0.35 mL/min.

Preparative HPLC separations were performed using an XTerra Prep RP_{18} OBD column (5 μm, 19×100 mm; Waters). The following mobile phases were used. Mobile phase #1 for preparative HPLC separation of peanut flower flavonoids: H_2O (A), CH_3CN (B), and 2% HCOOH in H_2O (C) were mixed in the following gradient: initial conditions, 65% A/ 30% B/ 5% C hold for 1 min, increase linearly to 0% A/ 95% B/ 5% C in 10 min, hold isocratic for 3 min, decrease to initial conditions in 0.01 min. The flow rate was 7.5 mL/min. Mobile phase #2 for preparative HPLC purification of dicoumaroylacetylspermidine: 74% H_2O (A), 24% CH_3CN (B), and 2% of 2%-HCOOH in H_2O (C). The flow rate was 8.5 mL/min. Mobile phase #3 for preparative HPLC purification of tricoumaroylspermidine: 70% H_2O (A), 28% CH_3CN (B), and 2% of 2%-HCOOH in H_2O (C). The flow rate was 9.0 mL/min.

Photoisomerization of Peanut Spermidine Conjugates 18 and 19. For analytical purposes, 0.5 to 20 mL of *ca.* 1×10^{-3} % solution of the conjugates in MeOH or mobile phases #2 and 3 were placed in polypropylene or borosilicate vials and irradiated in direct sunlight from 10 sec to 3 hours depending on the goal of the experiments. For preparative purposes, combined eluates (up to 250 mL) from the preparative HPLC column were placed in Pyrex flasks and irradiated in sunlight for 20 min. The experiments were performed at latitude 31.75° N in the middle of November from 11 am to 1 pm with clear sky; the intensity of the sunlight was not measured. Separation of the photoisomers was accomplished as described above for the natural spermidine conjugates using the same preparative HPLC column and mobile phase #2. Alternatively, for analytical purposes, a MeOH solution of spermidine conjugates was placed into 4-mL borosilicate vials equipped with lids with Teflon septa. The vials were placed in a horizontal position on a sheet of aluminum foil inside a UV viewing cabinet equipped with 4 × 15 W fluorescent lamps, and irradiated from a distance of 15 cm from 1 to 60 min. Photoisomerization at different temperatures (-95, 30, and 100 °C) was performed in a sealed ampoule, which was brought to a desired temperature under dim light and then exposed to sunlight for 15 min. To maintain temperature at 100 °C, the ampoule was placed in a beaker with boiling water (boiled with an electric heater). For a low-temperature experiment, the ampoule was placed in a beaker filled with MeOH, covered with a glass plate, and positioned in a Dewar flask with liquid N_2 so that most of MeOH became frozen; the temperature was monitored with a thermometer.

Data Analysis. Data were analyzed by the ANOVA method using SAS, 2000 (SAS Institute, Inc., Version 8, Cary, NC). Comparisons of the various means were performed by least significant difference (LSD) test at $p = 0.05$. Pearson product-moment correlation coefficient was calculated using SigmaStat (version 3.1, SYSTAT Software, Inc., Point Richmond, CA). Comparison of means of two groups of data was performed using the *t*-test. The Mann-Whitney Rank Sum Test was applied when the normality test failed ($p < 0.05$).

Spectroscopic measurements. UV measurements were performed with a Varian Cary 100 Bio UV-vis spectrophotometer. ^{1}H NMR, ^{13}C NMR, and homonuclear decoupling experiments were performed on a Bruker DRX-400 or a Bruker AC-300. HMBC and HMQC experiments were performed on a Bruker AMX-600 spectrometer. Chemical shift values were referenced to the solvent signals for CD_3OD (δ_H 3.31/δ_C 49.1) or DMSO-d_6 (δ_H 2.50/δ_C 39.5). HRESIMS and EIMS data were recorded on a Micromass Autospec instrument. ESI-MS/MS^2 data were obtained on a Finnigan LCQ Advantage MAX ion trap mass spectrometer equipped with an ESI interface and operated with Xcalibur™ version 1.4 software (Thermo Electron Corporation, San Jose, CA). The data were acquired in the full-scan mode from *m/z* 100 to 2000. Heated capillary temperature was 240°C, sheath gas flow 45 units, capillary voltage 10 V in positive mode and -4 V in negative, and source voltage 5.0 kV. In MS^2 analyses, the $[M+H]^+$ and $[M-H]^-$ ions observed for each chromatographic peak in full-scan analyses were isolated and subjected to source collision-induced dissociation (CID) using He buffer gas. In all CID analyses, the isolation width, relative fragmentation energy, relative activation Q, and activation time were *m/z* = 2.6, 20-45%, 0.25, and 30 msec, respectively. Concentrations of compound of interest were determined by reference to peak areas of corresponding pure compounds. The results of MS^2 experiments are represented throughout the text as follows: *m/z* aaa@bb → aaa, ccc, ddd, where aaa is parent ion, bb is normalized collision energy (%), and ccc and ddd are fragment ions.

UV spectra in mobile phase #1 for **13** - **16** are given in Figure 7. The UV spectra as well as MS and MS^2 data were nearly identical to those in the literature [116-118], with a characteristic loss of 162 daltons from the negatively charged pseudomolecular ions ($[M-H]^-$) of glucosides **13** and **15** (*m/z* 463 and 477, respectively) and a loss of 176 daltons from the pseudomolecular ions of glucuronides **14** and **16** (*m/z* 477 and 491, respectively).

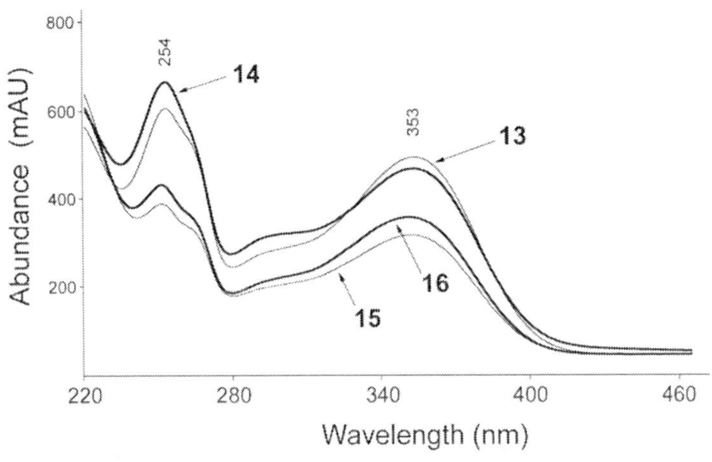

Figure 7. UV spectra (in mobile phase #1) of peanut flower flavonoid conjugates: 13 - quercetin-3-glucoside, 14 - quercetin-3-glucuronide, 15 - isorhamnetin-3-glucoside, and 16 - isorhamnetin-3-glucuronide.

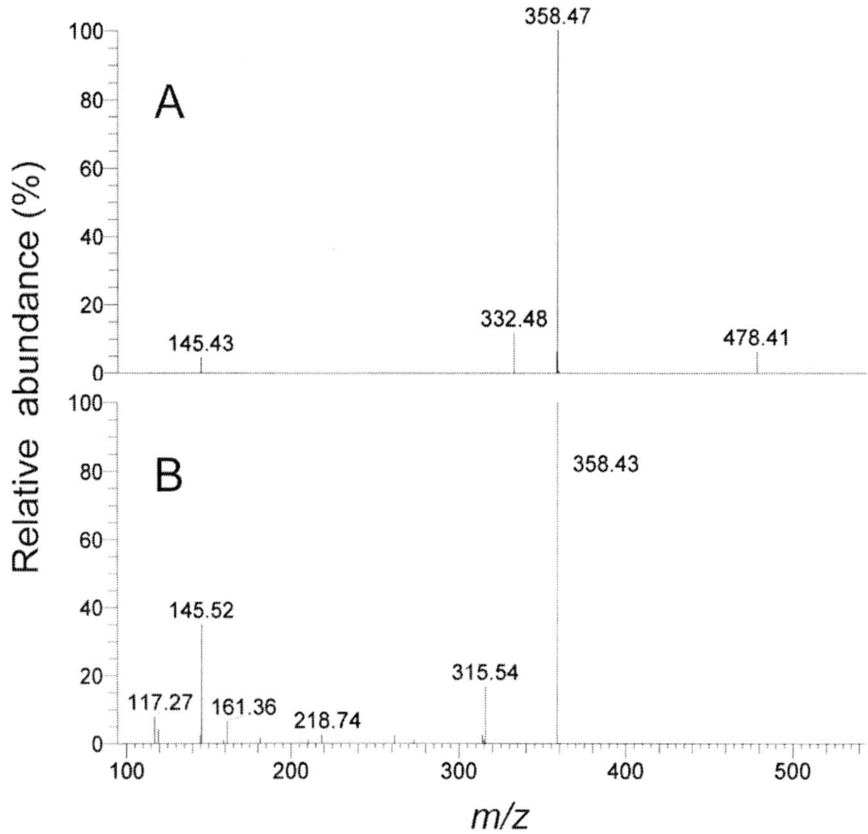

Figure 8. Negative ion ESI-MS spectra of *N*[1]-acetyl-*N*[5], *N*[10]-di-*p*-(*EE*)-coumaroylspermidine (**18**). A – *m/z* 478@38; B – *m/z* 478@38 → 358@45.

Di-*p*-(*EE*)-coumaroylspermidine (**17**). Negative ESI-MS^2 *m/z* 436@35: 436 ($[M\text{-}H]^-$; 65), 316 ($[M\text{-}H]^-$ - OH-C_6H_4-CH=CH; 100), 290 (1.2), 145 ($[M\text{-}H]^-$ - spermidine; 10). Positive ESI-MS^2 *m/z* 438@35: 438 ($[M+H]^+$; rel int 100), 422 (4.0), 292 (17), 274 (2.3), 204 (17), 147 (4.0). The MS data agree with those published in the literature [86].

N[1]-acetyl-*N*[5], *N*[10]-di-*p*-(*EE*)-coumaroylspermidine (**18**): white solid from CH_3CN and EtOAc; clear, very slightly greenish glass from MeOH. UV (mobile phase #2) λ_{max} (nm) 222, 292sh, 306. Negative ESI-MS (Figure 8): *m/z* 478@38: 478 ($[M\text{-}H]^-$; 7.1), 358 ($[M\text{-}H]^-$ - HO-C_6H_4-CH=CH; 100), 332 ($[M\text{-}H]^-$ - *p*-coumaroyl unit; 12), 145 ($[M\text{-}H]^-$ - spermidine; 4.9). Negative ESI-MS^3 (Figure 8) *m/z* 478@38 → 358@45: *m/z* 358 ($[M\text{-}H]^-$ - HO-C_6H_4-CH=CH; 100), 315 ($[M\text{-}H]^-$ - HO-C_6H_4-CH=CH - acetyl; 17), 145.5 ($[M\text{-}H]^-$ - spermidine; 35). Positive ESI-MS^2 *m/z* 480@35: 462 ($[M+H - H_2O]^+$; 2.2), 420 (11), 410 (3.3), 360 (4.0), 334 (89), 316 (100), 303 (46). 171 (8.6). EIMS (70 eV): molecular ion was not observed, but major fragment ions of potential significance were observed at *m/z* 333 (9), 293 (7), 265 (8), 247 (12), 186 (8), 181 (8), 169 (13), 147 (100), 129 (39), 119 (25), 100 (30), 84 (15), 70 (37), 69 (37); ^1H NMR at room temperature (400 MHz, DMSO-d_6): *p*-coumaroyl systems: δ 7.80-7.98 (m, NH-1 and NH-10), 7.26-7.51 (m, H-2, H-2', H-6, H-6', H-7 and H-7'), 6.70-6.85 (m, H-3,

H-3', H-5, H-5', H-8'), 6.32-6.38 (m, H-8): spermidine system: 3.00-3.50 (m, H_2-2, H_2-4, H_2-6, and H_2-9; partially obscured by water peak), 1.80 (br s, acetyl CH_3), 1.40-1.70 (m, H_2-3, H_2-7, and H_2-8). ^{1}H NMR at 55°C (400 MHz, DMSO-d_6): *p*-coumaroyl systems: δ 7.27-7.51 (m, H-2, H-2', H-6, H-6', H-7 and H-7'), 6.78 (d, J = 16, H-8 or H-8'), 6.70-6.76 (m, H-3, H-3', H-5, H-5'), 6.35 (d, J = 16, H-8 or H-8'): spermidine system: δ 7.80 (br s, NH-1 and NH-10), 3.00-3.50 (m, H_2-2, H_2-4, H_2-6, and H_2-9; partially obscured by water peak), 1.80 (s, acetyl CH_3) 1.40-1.70 (m, H_2-3, H_2-7, and H_2-8).

N^1-acetyl-N^5, N^{10}-di-*p*-(*ZZ*)-coumaroylspermidine (f, **Figure 9**): white solid from EtOAc. UV (mobile phase #2) λ_{max} (nm) 268. MS data were nearly identical to those of the *EE* isomer.

N^1-acetyl-N^5, N^{10}-di-*p*-(*ZE*)-coumaroylspermidine (g, **Figure 9**): white solid from EtOAc. UV (mobile phase #2) λ_{max} (nm) 217sh, 279sh, 285, 301sh. MS data were nearly identical to those of the *EE* isomer.

N^1-acetyl-N^5, N^{10}-di-*p*-(*EZ*)-coumaroylspermidine (h, **Figure 9**): white solid from EtOAc. UV (mobile phase #2) λ_{max} (nm) 225, 292, 305sh. MS data were nearly identical to those of the *EE* isomer.

N^1, N^5, N^{10}-tri-*p*-(*EEE*)-coumaroylspermidine (**19**). UV (mobile phase #3) λ_{max} (nm) 223, 292sh, 295, 304. HRESIMS *m/z* 584.2769 $[M + H]^+$, calcd for $C_{34}H_{37}N_3O_6$, 583.2684. Negative ESI- MS^2 *m/z* 582@35: 582 ($[M\text{-}H]^-$; 9.4), 462 ($[M\text{-}H]^-$ - OH-C_6H_4-CH=CH; 100), 436 ($[M\text{-}H]^-$ - *p*-coumaroyl unit; 8.4), 342 ($[M\text{-}H]^-$ - OH-C_6H_4-CH=CH - OH-C_6H_4-CH=CH; 3.2). Positive ESI-MS^2 *m/z* 584@35: 566 ($[M\text{+}H\text{-}H_2O]^+$; 2.1), 464 (5.6), 438 (70), 420 (100), 275 (1.2), 204 (0.5). ^{1}H NMR (room temperature, 400 MHz, CD_3OD): *p*-coumaroyl systems: δ 7.49-7.55 (m, H-7 and H-7'), 7.33-7.48 (m, H-2, H-2', H-2", H-6, H-6', H-6" and H-7"), 6.85 (m, H-8"), 6.70-6.83 (m, H-3, H-3', H-3", H-5, H-5' and H-5"), 6.34-6.43 (m, H-8 and H-8'): spermidine system: δ 3.32-3.60 (m, H_2-2, H_2-4, H_2-6, and H_2-9), 1.54-1.96 (m, H_2-3, H_2-7, and H_2-8). ^{1}H NMR (room temperature, 400 MHz, DMSO-d_6): *p*-coumaroyl systems: δ 7.27-7.54 (m, H-2, H-2', H-2", H-6, H-6', H-6", H-7, H-7' and H-7"), 6.85 (m, H-8"), 6.67-6.77 (m, H-3, H-3', H-3", H-5, H-5' and H-5"), 6.33-6.43 (m, H-8 and H-8'): spermidine system: δ 7.94-8.10 (br m, NH-1 and NH-10), 3.12-3.50 (m, H_2-2, H_2-4, H_2-6, H_2-9), 1.40-1.80 (m, H_2-3, H_2-7, H_2-8). ^{1}H NMR at 95°C (400 MHz, DMSO-d_6) *p*-coumaroyl systems: δ 7.28-7.46 (m, H-2, H-2', H-2", H-6, H-6', H-6", H-7, H-7' and H-7"), 6.80 (d, J = 16, H-8"), 6.73-6.78 (m, H-3, H-3', H-3", H-5, H-5' and H-5"), 6.37 (d, J = 16, H-8 or H-8'), 6.38 (d, J = 16, H-8 or H-8'): spermidine system: δ 7.56-7.68 (br m, NH-1 and NH-10), 3.12-3.50 (m, H_2-2, H_2-4, H_2-6, H_2-9), 1.40-1.80 (m, H_2-3, H_2-7, H_2-8). The MS and UV data agree with those published in the literature [86, 88, 104].

The UV spectra of both **18** and **19** did not change in acidic (pH 4.0), neutral, or basic (pH 8.5) solutions. Neither of compounds steam-distill. As expected, irradiation of **18** and **19** with green light (532 nm) from a 5-mW laser pointer for 40 min did not cause any change in the compounds, which indicates that green-light illumination may be used during the isolation of the compounds.

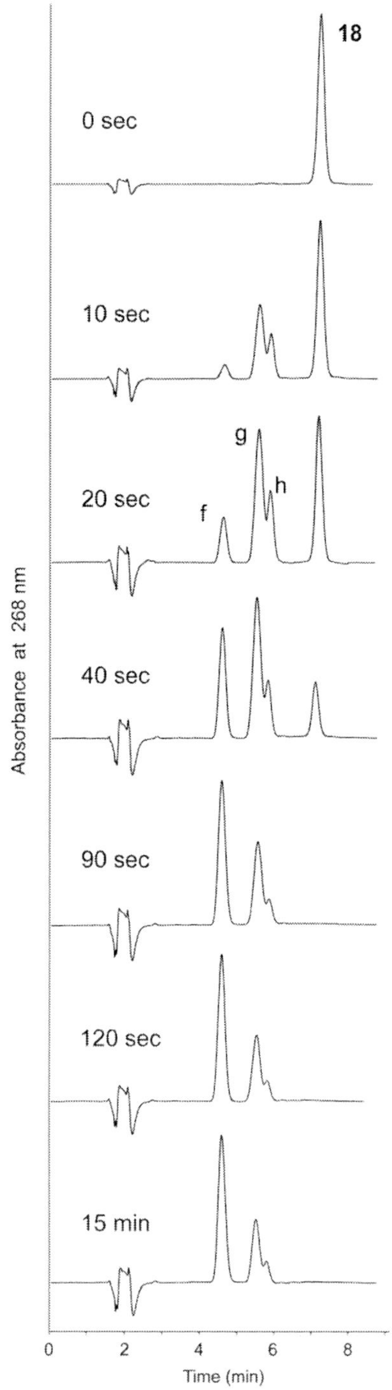

Figure 9. Dynamics of N^1-acetyl-N^5, N^{10}-di-*p*-(*EE*)-coumaroylspermidine (**18**) photoisomerization. c - N^1-acetyl-N^5, N^{10}-di-*p*-(*EE*)-coumaroylspermidine; d and e –(*EZ*)- and (*ZE*)-di-*p*-coumaroylacetylspermidine isomers. Chromatograms were normalized by the highest peak.

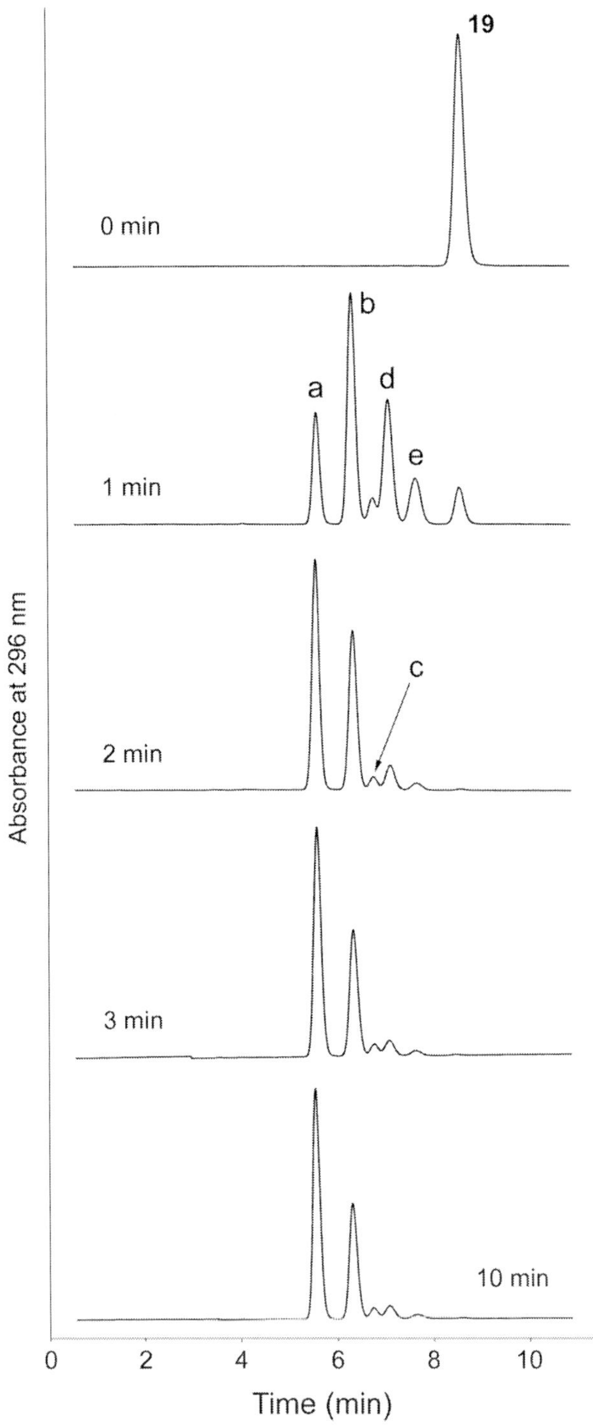

Figure 10. Dynamics of N^1, N^5, N^{10}-tri-*p*-(*EEE*)-coumaroylspermidine (**19**) photoisomerization. a - N^1, N^5, N^{10}-tri-*p*-(*ZZZ*)-coumaroylspermidine; b – mixture of *ZZE* isomers of N^1, N^5, N^{10}-tri-*p*-coumaroylspermidine. Chromatograms were normalized by the highest peak.

N^1, N^5, N^{10}-tri-*p*-(*ZZZ*)-coumaroylspermidine (a, Figure 10). UV (mobile phase #3) λ_{max} (nm) 268. MS data were nearly identical to those of the *EEE* isomer. ^{1}H NMR (400MHz, CD_3OD): *p*-coumaroyl systems: δ 7.34-7.42 (m, H-2' and H-6'), 7.18-7.23 (m, H-2" and H-6"), 6.69-6.75 (m, H-3', H-3", H-5' and H-5"), 6.50-6.65 (6 d, each with *J* = 13, H-7, H-7' and H-7") 5.75-5.92 (six doublets, each with *J* = 13, H-8, H-8', and H-8" signals for both conformers): spermidine system: 3.10-3.40 (m, H_2-2, H_2-4, H_2-6, H_2-9), 1.30-1.80 (m, H_2-3, H_2-7, H_2-8).

Mixture of *ZZE* isomers of N^1, N^5, N^{10}-tri-*p*-coumaroylspermidine (b, Figure 10). UV (mobile phase #3) λ_{max} (nm) 279, 302sh. MS data were nearly identical to those of the *EEE* isomer.

Results and Discussion

Structure Elucidation. Three major peaks (13, 14, and 19; Figure 11) were detected in the flower extracts. Two of the peaks (13 and 14) were suggested to be flavonoid conjugates based on their characteristic absorption (Figure 7) and mass spectra. Inspection of the ^{1}H and ^{13}C NMR spectra for compounds **13** and **14** suggested that both compounds are flavonoid glycosides. In each instance, analysis of the data revealed the presence of a 1,3,5-trioxygenated 1,2,3,5-tetrasubstituted benzene ring and a 1,2-dioxygenated 1,2,4-trisubstituted benzene ring, as well as a six-carbon monosaccharide unit and three other sp^2 carbons indicating the presence of a monoglycosylated quercetin analogue. The signals for the monosaccharide unit in compound **13** corresponded to those of a glucuronic acid unit, while those of compound **14** were consistent with the presence of a glucose moiety. ^{1}H NMR *J*-values enabled assignment of a β-linkage in both cases. For both compounds, the signal for the anomeric proton showed an HMBC correlation to C-3 of the γ-pyrone ring, locating the monosaccharide unit at the 3-position of the flavonoid. Detailed analysis of the HMBC data were fully consistent with identification of these compounds as quercetin-3-glucoside, and quercetin-3-glucuronide (13 and 14; Figure 5).

Compounds **15** and **16** were identified as isorhamnetin-3-glucoside and isorhamnetin-3-glucuronide, respectively, on the basis of MS and UV data. The compounds were not obtained in quantities conducive to independent structure determination by NMR analysis, so the positions of the glucosyl and glucuronyl moieties were not unambiguously confirmed. However, structures **15** and **16** are considered most likely since the major compounds **13** and **14** bear 3-glucosyl and 3-glucuronyl units, respectively (Figure 5). In addition, MS and UV data for **15** and **16** were virtually identical to those published in the literature [116-118]. Compounds **13** and **14** gave $[M-H]^-$ ions at *m/z* 463 and *m/z* 477, respectively. Upon MS^2 analysis, each produced a major fragment ion at *m/z* 301 ($[M-H]^-$ -162 daltons in **13**, and $[M-H]^-$ -176 daltons in **14**, corresponding to loss of the glucose and glucuronide units, respectively). Compounds **15** and **16** afforded $[M-H]^-$ ions at *m/z* 477 and *m/z* 491, respectively, both of which fragmented by MS^2 to afford a major ion at *m/z* 315 ($[M-H]^-$ -162 daltons in **15**; and $[M-H]^-$ -176 daltons in **16**, again corresponding to loss of the glucose and glucuronide units, respectively).

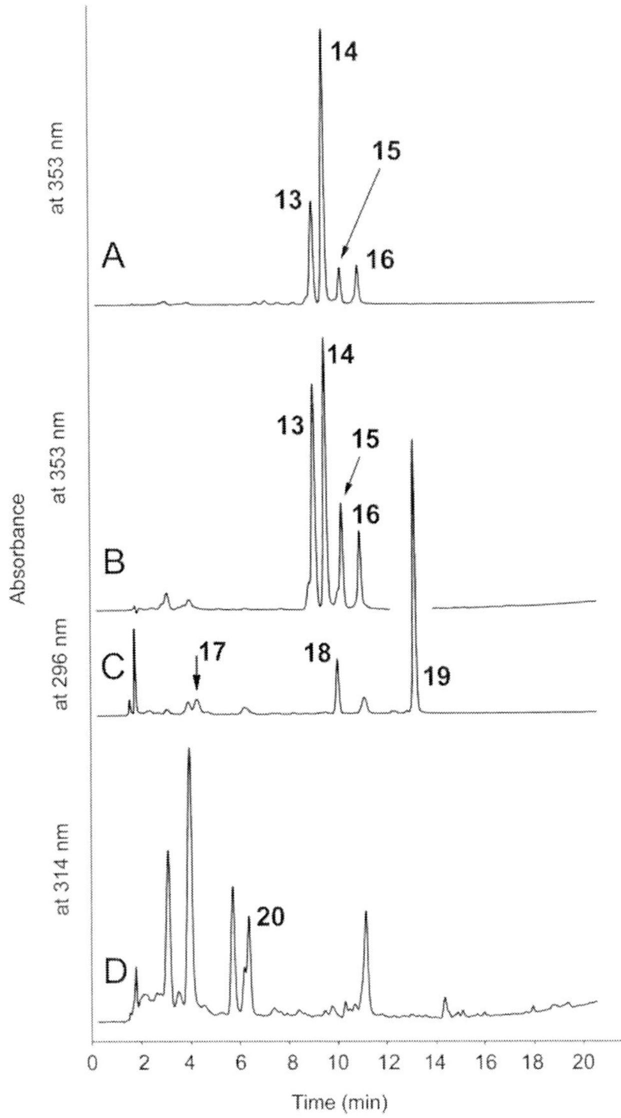

Figure 11. Flavonoid and spermidine conjugates content in different parts of the peanut flower. A - standard; B – wings; C – keel; D – hypanthium. 13 - Quercetin-3- glucoside; 14 - Quercetin-3-glucuronide; 15 - Isorhamnetin-3- glucoside; 16 - Isorhamnetin-3- glucuronide; 17 – di-*p*-(*EE*)-coumaroylspermidine; 18 - N^1-acetyl-N^5, N^{10}-di-*p*-(*EE*)-coumaroylspermidine; 19 - N^1, N^5, N^{10}-tri-*p*-(*EEE*)-coumaroylspermidine; 20 - *p*-(*E*)-coumaric acid. Chromatograms were normalized by the highest peak.

The UV spectrum of the third major peak (19) resembled those typical of *p*-hydroxycinnamic acids, but the compound could not be clearly identified based only on its UV and MS data. A database search for the accurate mass value of *m/z* 584.2769 obtained by HRESIMS for the $[M + H]^+$ ion of compound **19** suggested that the compound is likely a trisubstituted spermidine conjugate wherein each spermidine nitrogen atom is incorporated into an amide linkage with a hydroxycinnamic acid-type unit. Detailed analysis of NMR data revealed the presence of a complex mixture of overlapping signals that could not be readily analyzed. However, the mass and the general appearance of the spectrum were both consistent with a tri-*p*-coumaroylspermidine derivative. The complexity of the spectrum was attributed to the presence of a mixture of conformers presumably due to the presence of a tertiary amide unit, and was consistent with literature reports [86, 88, 94]. NMR data were collected at elevated temperature in an effort to accelerate interconversion of the conformers, thereby simplifying the NMR data. As anticipated, coalescence of signals was observed upon increasing the temperature of a sample dissolved in DMSO-d_6 (Figure 12). At 95 °C, a single set of sharp signals was observed, enabling more straightforward interpretation of the data. Clear observation of individual doublets corresponding to the upfield olefinic protons of the *p*-coumaroyl units in this spectrum enabled observation of a 16-Hz *J*-value for each signal, resulting in assignment of the *E*-geometry for all three olefin units.

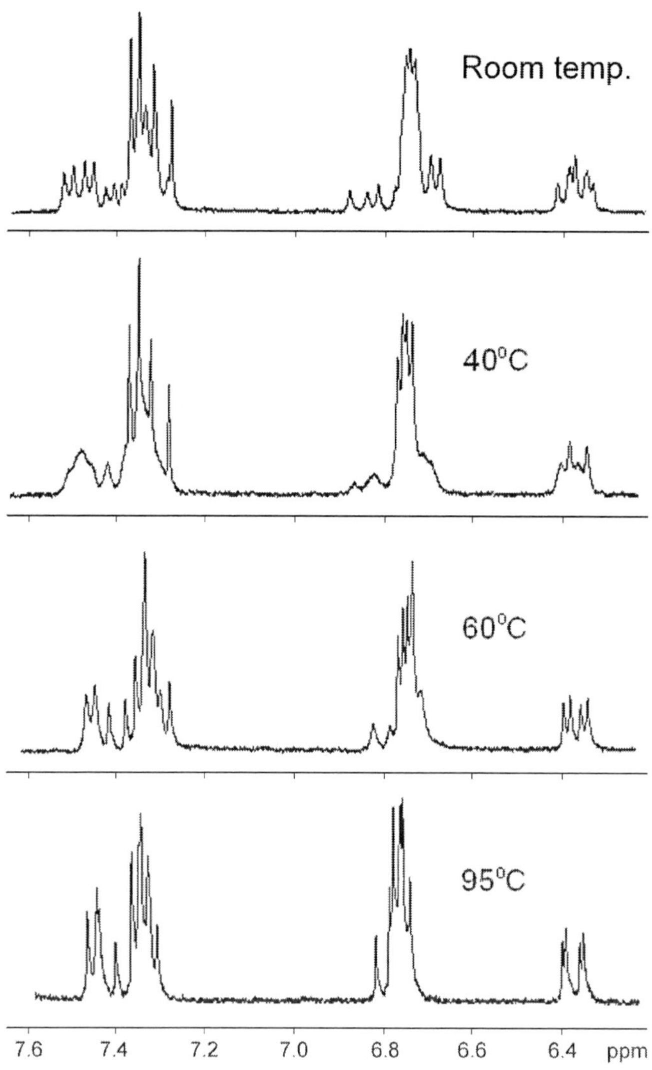

Figure 12. Aromatic/olefinic region of the ^{1}H NMR spectrum of N^1, N^5, N^{10}-tri-*p*-(*EEE*)-coumaroylspermidine (**19**) showing signal coalescence at elevated temperature (DMSO-d_6).

The ESI mass spectrum of **18** (Figure 8) revealed a molecular weight of 477 daltons, and MS^2 data showed a loss of 43 daltons from the major daughter ion at *m/z* 358. These results suggested the replacement of one of the *p*-coumaroyl groups found in **19** with an acetyl group in **18**. The ^{1}H NMR data were similar to those of **19**, again showing a complex mixture of signals expected for a mixture of conformers, but were somewhat simpler, and also contained a broad singlet characteristic of the acetyl unit of the conformers. As was the case for **19**, ^{1}H NMR analysis at elevated temperature afforded a simplified spectrum that was fully consistent with assignment of **18** as a di-*p*-(*EE*)-coumaroylacetylspermidine. However, the question of regiochemistry is a challenging one. The relative positions of the *p*-coumaroyl and acetyl moieties in **18** could not be deduced from NMR data due to overlap of the methylene signals that occurs even at high field. Similarly, positional isomers of other triacylated spermidine derivatives (coumaroyldicaffeoylspermidine and dicoumaroylcaffeoylspermidine) could not be distinguished on the basis of NMR data [88]. In the ^{1}H NMR spectrum of **19** (at elevated temperature), it was noted that two of the olefinic signals corresponding to the 8-positions of two of the *p*-coumaroyl units were significantly upfield of the other (δ 6.37 and 6.38 vs. 6.80). Given the close similarity in the environments of the two terminal *p*-coumaroyl units (at N-1 and N-10), the most downfield signal of the three was presumed to arise from the corresponding proton in the central *p*-coumaroyl unit (the one at N-5). In the ^{1}H NMR data for **18**, also at elevated temperature, the two *p*-coumaroyl position-8 doublets appeared at δ 6.80 and 6.38, suggesting that one of the terminal *p*-coumaroyl units was missing, and thereby suggesting the placement of the acetyl group at a terminus (N-1 or N-10). Even if this conclusion is correct, no decision could be made on the basis of the available NMR data regarding whether the acetyl group was at the N-1 or N-10 position. Evaluation of the EIMS fragmentation pattern for **18** was considered in an effort to locate the acetyl unit, although literature reports indicate that transamidation may lead to ambiguous conclusions. For example, Bokern [88], Bigler [119], and Hu [120] noted the occurrence of a transamidation (the *Zip* reaction) of the $[M+H]^+$ ions for dicoumaroylspermidine under MS conditions. Even so, the EIMS fragmentation pattern enabled a proposal for the location. The base peak in the spectrum (*m/z* 147) corresponded to the *p*-coumaroyl acylium ion. The next most abundant peak (*m/z* 129) suggested that the acetyl group is located at the N-1 position. Given the corresponding substitution pattern, such an ion could be rationalized by α-cleavage to the N-5 position (between C6 and C7 of the spermidine unit), accompanied by elimination of the N-5 coumaroyl unit. Neither of the other two regioisomers would be expected to give such an ion in a straightforward way. The ion corresponding to the analogous process beginning with α-cleavage on the opposite side of N-5 (*m/z* 247) was also observed at moderate intensity. The EI mass spectrum of triacetylspermidine shows the analogous fragmentations as the two most abundant peaks in the spectrum (*m/z* 129 and 143) [121]. On the basis of these considerations, and the fact that the suggestions provided by the NMR and MS data are consistent with each other, the acetyl group of 18 is proposed to be at the N-1 position. However, it was ultimately felt that the data described here do not permit a truly conclusive assignment.

The structure of **17** was deduced from UV and MS data, which matched those previously published [86], although the structure assignment in this instance was not considered unambiguous due to insufficient quantities available for analysis by NMR.

Dynamics of Photoisomerization.

Interestingly, spermidine conjugates **18** and **19** exhibited very high sensitivity to sunlight. The sample composition (as judged by HPLC profile) changed dramatically and rapidly upon exposure to sunlight for only a very short time. Similar changes were observed upon irradiation with UV light, but required longer reaction times. Since sunlight is clearly a natural environmental factor for the peanut plant, efforts to explore this phenomenon were undertaken.

Samples were irradiated with ambient sunlight or UV light, resulting in photoisomerization of the olefin units from the *E* to *Z* forms. Such a transformation is not unexpected for such compounds based on literature precedents [86, 94, 122], but the facility and the extent of the conversion was somewhat unexpected. Due to the asymmetry of N^1, N^5, N^{10}-tri-*p*-coumaroylspermidine (**19**) molecule [-$(CH_2)_3$- between N^1 and N^5] and [-$(CH_2)_4$- between N^5 and N^{10}], eight stereoisomers having various geometries of the three existing olefinic double bonds are possible: *EEE*, *EEZ*, *EZE*, *EZZ*, *ZEE*, *ZZE*, *ZEZ*, and *ZZZ*. Six chromatographic peaks, including those corresponding to the *EEE* isomer originally present, were observed after only very brief exposure to light (Figure 10, 1 min; Figure 13). After longer exposure, the peak for the original *EEE* isomer disappeared and the system came to an equilibrium that was observed in the chromatogram as two major and three minor peaks (Figure 10, 1-10 min). LCMS data indicated that all of the peaks retained the same molecular weight, as expected for a mixture of isomerization products. Samples corresponding to the two major product peaks (peaks a and b, Figure 10) were collected by HPLC and analyzed by ^{1}H NMR. Both samples showed the same kind of signal complexity associated with the conformational mixtures observed at room temperature. However, the data for the major component were sufficiently resolved to enable a determination that the olefin *J*-values had all changed to 13 Hz, indicating that all of the olefins in this product now bear the *Z*-configuration. The greater complexity of the NMR data for the other major peak was presumed to be due to the presence of a mixture of isomers that included two *Z*-olefin units and one E-olefin units, and this was consistent with observations of both 13-and 16-Hz *J*-values for individual signals appearing in the olefinic region.

For the dicoumaroylacetylspermidine derivative (**18**), only four isomers are possible (*EE*, *EZ*, *ZE*, and *ZZ*). After brief exposure of a sample of pure **18** to sunlight, four peaks, including the original *EE* isomer were present in the chromatogram (Figure 9, 20 sec; Figure 14). However, by the end of a 90-sec exposure, only two major and one minor peaks were detected, and the original *EE* isomer was absent (Figure 9, 90 sec). Continued irradiation for 15 min did not cause further change in the composition or ratio of the components of the mixture (Figure 9, 15 min).

Separation of the photoisomers could be achieved on a C_{18} stationary phase only when CH_3CN was used as an organic modifier of the mobile phase. Isocratic elution was found to be superior to gradient elution and allowed for baseline separations of the isomers. The ratios of isomers obtained in the products at equilibrium were judged to be reflective of approximately a 6:1 preference for *Z* over *E* at each possible olefin position, which is reasonably consistent with expectations based on literature precedents for photoisomerization of similar compounds [86, 122]. Analogous results were obtained regardless of temperature, as the same chromatographic results were obtained over similar time-periods when the experiments were conducted at -95, 30, and 100 °C. Variation in the solvent employed (MeOH or CH_3CN-H_2O-HCOOH) also led to no change in the results.

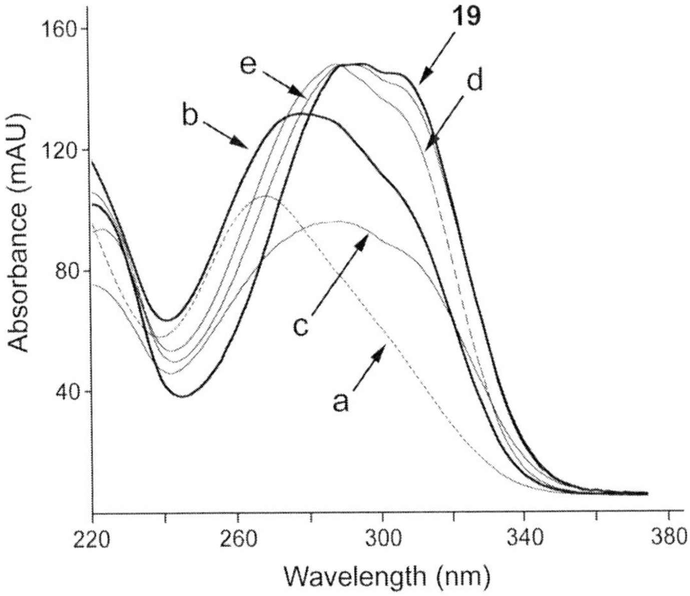

Figure 13. UV spectra of N^1, N^5, N^{10}-tri-*p*-(*EEE*)-coumaroylspermidine (**19**) and its photoisomers: a - N^1, N^5, N^{10}-tri-*p*-(*ZZZ*)-coumaroylspermidine; b – mixture of *ZZE* isomers of N^1, N^5, N^{10}-tri-*p*-coumaroylspermidine, c and d - mixture of *ZZE*/*EEZ* isomers, e – mixture of *EEZ* isomers.

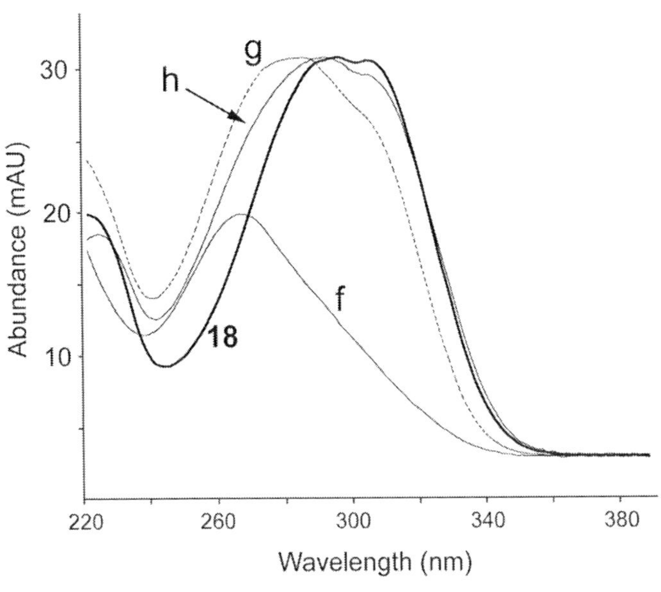

Figure 14. UV spectra of N^1-acetyl-N^5, N^{10}-di-*p*-(*EE*)-coumaroylspermidine (**18**) and its photoisomers: f - N^1-acetyl-N^5, N^{10}-di-*p*-(ZZ)-coumaroylspermidine; g and h mixture of *EZ*, *ZE* isomers.

Production of Major Secondary Metabolites

Due to lack of the data in literature on peanut flower constituents, the major low molecular weight secondary metabolites in the whole flower extracts were investigated first. HPLC chromatograms of these extracts revealed the presence of three major components (**13**, **14**, and **19**; Figure 11). Extracts of individual peanut flower organs (Figure 4) contained diverse major compounds. The standard and wings contained high concentrations of **13**, **14**, **15**, and **16** (Figure 11, Table 1). The wings tended to accumulate **13** and **14** at a lower ratio to **15** and **16** than the standard. Flavonoid conjugates **13 - 16** were virtually the only metabolites that were extracted with MeOH. These compounds are commonly found in other plants (83, 117], but **13-16** are reported in peanut flowers for the first time in the present study. On the contrary, flavonoids **13-16** were not detected in the keel, which was instead found to accumulate spermidine conjugates **17**, **18**, and **19**. The latter compound was found in high concentrations (Table 1). Quantitation of **17** in the whole flower extracts could not be accomplished due to its low concentrations. The hypanthium extract was comprised mainly of simple phenolic acids, of which *p*-coumaric acid is most notable given its involvement in the formation of spermidine conjugates **17**, **18**, and **19**. Concentrations of *p*-coumaric acid in the hypanthium are reflected in Table 1.

Although there was a significant difference in flower weight among some breeding lines (Table 1), there was no correlation between the weight (and, indirectly the number of flowers) and flavonoid and spermidine conjugate production (Table 1). Side-by-side comparison of production of the same compounds by the two market-type peanut varieties represented in Table 1 did not reveal any significant difference. Production of flavonoids and spermidines was uniform, but there was significant difference between some individual breeding lines of both market types. For example, production of *p*-coumaric acid by Tifrunner was the highest of all tested breeding lines. At the same time, overall concentration of the flavonoid conjugates **13-16** was the lowest in this line, and there was a significant difference between production of **13** and **14** by Tifrunner and other representatives of the runner type peanuts. In some breeding lines, only two metabolites (**14** and *p*-coumaric acid) varied significantly in Virginia market type peanuts; Gregory and experimental line B accumulated these metabolites in about twice the concentration compared with AT3081 R and experimental line B. At the same time, concentrations of **19** did not vary significantly among all tested breeding lines. It seems that the biosynthesis of **19** by the flower reproductive organs is a well-regulated process that does not depend on peanut genotypes (Table 1).

In the present study, compounds **18** and **19** demonstrated very high sensitivity to sunlight (Figures 9 and 10). It should be emphasized that only *E*-isomers of the spermidine conjugates **17**, **18**, and **19** were detected in peanut flowers when extracts and solutions were protected from light. The configurational *E*/*Z* isomeric mixtures were found only during isolation of the products and in other experiments when protection from light was not assured. It should also be noted that coumaroylacetylspermidine **18** is not an artifact since it was detected in high concentrations under all extraction conditions, including soaking fresh whole flowers in MeOH at 4 ^{o}C.

Table 1. Major constituents of whole peanut flowers (μg/g fresh[1] tissue).

Peanut breeding line	Flower weight (g)	*p*-Coumaric acid	Quercetin-3-Glc[2] (**13**)	Quercetin-3-GlcUA[2] (**14**)	Isorhamnetin-3-Glc[2] (**15**)	Isorhamnetin-3-GlcUA[2] (**16**)	DCAS[3]	TCS[4]
Runner market type								
AT29-1112	179.5	12.5±2.2	66.2±5.1	76.0±4.2	10.1±2.3	13.9±1.8	8.7±1.8	42.3±3.2
G02C	122.6	14.6±1.3	40.7±3.8	89.0±3.5	7.6±1.1	14.8±1.5	6.8±1.1	37.6±2.0
AT3-1114	104.8	15.1±1.8	38.5±1.9	78.1±4.1	13.8±1.5	12.6±1.2	10.1±1.6	59.6±2.5
C724-19-15	90.5	22.0±2.1	44.5±3.5	83.6±3.8	9.9±0.8	18.3±2.1	8.4±0.8	46.0±1.9
Tifrunner	139.6	23.1±2.4	31.5±2.9	58.1±4.3	11.4±1.0	18.8±2.0	10.6±1.2	56.1±2.4
Virginia market type								
AT3081 R	132.8	6.9±0.7	44.6±3.0	41.2±3.3	6.8±1.4	10.6±1.4	9.2±1.3	44.4±2.7
Exp. line A	221.9	8.5±0.9	51.6±3.7	47.2±3.9	8.6±1.8	12.4±2.1	9.2±1.4	42.1±1.8
VC-2	158.7	9.3±1.6	42.3±2.6	85.7±4.1	13.6±2.2	12.4±2.5	7.8±1.2	41.0±2.8
AT3085 RO	162.8	12.2±1.8	44.5±3.1	52.6±3.4	11.1±1.2	12.0±2.0	10.9±2.1	48.8±2.4
Gregory	108.2	13.6±1.8	42.9±3.2	84.1±3.6	7.5±0.7	13.6±1.7	9.5±0.9	51.2±2.2
Exp. line B	180.2	16.9±2.5	39.2±2.4	87.9±4.6	8.5±1.4	15.1±2.6	6.7±1.0	35.5±3.0

[1] A multiplication factor of 9.3 should be used for recalculation of the major constituents' content in dry weight of plant material.

[2] GlcUA = glucuronyl unit, Glc = glucosyl unit; [3] DCAS = N^1-acetyl-N^5, N^{10}-di-*p*-(*EE*)-coumaroylspermidine (**18**); [4] TCS = N^1, N^5, N^{10}-tri-*p*-(*EEE*)-coumaroylspermidine (**19**).

Earlier experiments [109] demonstrated that in strawberry, a short-day plant, spermidine conjugates along with other free and bound polyamines were accumulated in the shoot tips during floral induction and before floral emergence. In vegetative shoot tips from plants grown continuously during long days, conjugated spermidines as well as free and bound amines did not appear; free and bound polyamine concentrations were low and no change was observed. Male-sterile flowers were distinguished by their lack of conjugated spermidines and free and bound amines. Apparently, light directly or indirectly regulates conjugated spermidine production/accumulation, but the mechanism of this phenomenon is unknown.

The peanut flower reproductive organs are located in a curved keel (Figure 4) that is formed by two petals fused along the dorsal edges to the apex, but open ventrally at the base [69]. The keel seems to be well protected by the petals from the UV portion of the sunlight taking into account the high concentrations of the flavonoids and presence of carotenoids. However, dormant flowers may be exposed to sunlight for several hours during daytime, and in this unlikely scenario, the chemical shield may not necessarily be able to completely block the radiation that may be absorbed by the spermidines. If so, formation of the spermidine photoisomers (Figures 9 and 10) is possible, considering the extremely high photosensitivity of the major spermidine conjugates **19** (Figure 10) and **18** (Figure 9). However, it is also possible that the flower reproductive organs may have some way to maintain spermidines **17** - **19** in their natural *E* configuration. In the present study, the photoisomers were not detected, as the flowers were collected at sunrise, at the time when fertilization occurs. Additional experiments with dormant flowers may reveal the presence or absence of the photoisomers. Whether photoisomerization may take place in peanut flowers in the daytime, and whether the photoisomers may play a role in one or more biochemical processes regulating plant reproduction, are questions that remain to be addressed.

CONCLUSION

A great range of morphological and reproductive diversity exists in the genus *Arachis*, an old group of unusual plants. A number of the morphological features of peanut associated with pest and disease resistance are agronomically and economically important. Recent research on chemical and physiological aspects of peanut plant growth and development has revealed new facts about the nature of this plant. The present study revealed the production of flavonoid (**13-16**) and spermidine conjugates (**18** and **19**) in the peanut flower at high levels. All of these major metabolites were detected in the flowers for the first time. One of the metabolites, N^1-acetyl-N^5, N^{10}-di-*p*-(*EE*)-coumaroylspermidine (**18**), has not been previously reported in plants. Studies of the changes that occur in **18** and **19** upon irradiation with sunlight revealed formation of several intermediate photoisomers. Flavonoid and spermidine conjugates may play protective roles against pests, as reflected in the cited literature. The exact role of **18** is unknown, but as has been suggested for **19**, the compound may be involved in regulating the plant reproduction process.

The authors of the present research are planning to study biological activity of the new spermidine derivative (**18**), which is so far unique to the peanut flower. Future research will also concentrate on studies of potential low molecular weight chemical markers, similar to the spermidines and stilbenoids, that may help in breeding pest-resistant cultivars with uncompromised vigor. Collaboration with peanut breeders is foreseen. At present, breeding is the most efficient single technique that allows generation of new prospective peanut varieties. In the past, cross-incompatability factors have limited introduction of favorable genes from wild *Arachis* spp. into cultivated peanuts. However, this may be circumvented through tissue culture techniques and genetic engineering. Further knowledge of peanut chemistry and physiology may prove to be very valuable in leading to the development of improved peanut cultivars..

Abbreviations Used

DAD	–	Diode Array Detector
HPLC	–	High Performance Liquid Chromatography
MS	–	mass spectrometer, mass spectrometry, mass spectra
ESI	–	Electrospray Ionization
HRESIMS	–	High Resolution Electrospray Ionization Mass Spectrometry
EIMS	–	Electron Impact Mass Spectrometry
HMBC	–	Heteronuclear Multiple Bond Correlation
HMQC	–	Heteronuclear Multiple Quantum Correlation

References

[1] Dwivedi, SL. *Molecular diversity in Arachis hypogaea germplasm.* 2008/10/23. Available from: URL: http://test1.icrisat.org/gt-bt/ResearchBriefs/mdingn.html

[2] *World Geography of the Peanut.* 2008/10/23. Available from: URL: http://www.lanra.uga.edu/peanut/knowledgebase

[3] Stalker, HT; Simpson, CE. (1995). Germplasm resources in Arachis. In HE; Pattee, HT. Stalker, (Eds.), *Advances in peanut science*, (286-300). Stillwater, OK. American Peanut Research and Education Society, Inc.

[4] *Peanut Facts.* 2008/10/23. Available from: URL: http://www.soyatech.com/peanut_facts.htm

[5] *George Washington Carver.* 2008/10/23. Available from: URL: http://www.ideafinder.com/ history/inventors/carver.htm

[6] *US Patents*: 5188064, 5361719, 5526770.

[7] Sobolev, VS; Cole, RJ. Note on utilization of peanut seed testa. *J. Sci. Food Agric.*, 2004, 84, 105-111.

[8] *Medical Uses of Pycnogenol.* 2008/10/23. Available from: URL: http://www.diet-and-health.net/Supplements/Pycnogenol.html

[9] *US Patent* 4,698,360

[10] Sobolev, VS; Cole, RJ. trans-Resveratrol content in commercial peanuts and peanut products. *J. Agric. Food Chem.*, 1999, 47, 1435-1439.

[11] Fritzemeier, KH; Rolfs, CH; Pfau, J; Kindl, H. Action of ultraviolet-C on stilbene formation in callus of Arachis hypogaea, *Planta*, 1983, 159, 25-29.

[12] Sobolev, VS; Deyrup, ST; Gloer, JB. New peanut (Arachis hypogaea) phytoalexin with stilbene and but-2-enolide moieties. *J. Agric. Food Chem.*, 2006, 54, 2111-2115.
[13] Sobolev, VS; Horn, BW; Potter, TL; Deyrup, ST; Gloer, JB. Production of stilbenoids and phenolic acids by the peanut plant at early stages of growth. *J. Agric. Food Chem.*, 2006, 54, 3505-3511.
[14] Paxton, JD. (1991). Biosynthesis and accumulation of legume phytoalexins. In RP; Sharma, & DK. Salunkhe, (Eds.), *Mycotoxins and Phytoalexins*, (485-499). Boca Raton, FL: CRC Press..
[15] Subba Rao, PV; Strange, RN. (1995). Chemistry, Biology, and Role of Groundnut Phytoalexins in Resistance to Fungal Attack. In M; Daniel, & RP. Purkayastha, (Eds.), *Handbook of Phytoalexin Metabolism and Action*, (199-227). New York, NY: Marcel Dekker, Inc.
[16] Cole, RJ; Cox, RH. (1981). *Handbook of Toxic Fungal Metabolites*. New York, NY: Academic Press.
[17] Arora, MK; Strange, RN. Phytoalexin accumulation in groundnuts in response to wounding. *Plant Sci.*, 1991, 78, 157-163.
[18] Cooksey, CJ; Garratt, PJ; Richards, SE; Strange, RN. A dienyl stilbene phytoalexin from Arachis hypogaea. *Phytochemistry*, 1988, 27, 115-116.
[19] Aguamah, GA; Langcake, P; Leworthy, DP; Page, JA; Pryce, RJ; Strange, RN. Two novel stilbene phytoalexins from Arachis hypogaea. *Phytochemistry*, 1981, 20, 1381-1383.
[20] Keen, NT; Ingham, JL. New stilbene phytoalexins from American cultivars of Arachis hypogaea. *Phytochemistry*, 1976, 15, 1794-1795.
[21] Wotton, HR; Strange, RN. Circumstantial evidence for phytoalexin involvement in the resistance of peanuts to Aspergillus flavus. *J. Gen. Microbiol*, 1985, 131, 487-494.
[22] Sobolev, VS; Cole, RJ; Dorner, JW; Yagen, B. Isolation, purification, and liquid chromatographic determination of stilbene phytoalexins in peanuts. *J. AOAC Int.*, 1995, 78, 1177-1182.
[23] Sobolev, VS; Potter, TL; Horn, BW. Prenylated stilbenes from peanut root mucilage. *Phytochem Anal*, 2006, 17, 312-322.
[24] Wotton, HR; Strange, RN. Increased susceptibility and reduced phytoalexin accumulation in drought-stressed peanut kernels challenged with Aspergillus flavus. *Appl. Environ. Microbiol*, 1987, 53, 270-273.
[25] Dorner, JW; Cole, RJ; Sanders, TH; Blankenship, PD. Interrelationship of kernel water activity, soil temperature, maturity, and phytoalexin production in preharvest aflatoxin contamination of drought-stressed peanuts. *Mycopathologia*, 1989, 105, 117-128.
[26] Ingham, JL. 3,5,4'-Trihydroxystilbene as a phytoalexin from groundnuts (Arachis hypogaea). *Phytochemistry*, 1976, 15, 1791-1793.
[27] Sobolev, VS; Neff, SA; Gloer, JB. New Stilbenoids from Peanut (Arachis hypogaea) Seeds Challenged by an Aspergillus caelatus Strain. *J. Agric. Food Chem.*, 2009, 57, 62-68.
[28] Keen, NT. The isolation of phytoalexins from germinating seeds of Cicer arietinum, Vigna sinensis, Arachis hypogaea, and other plants. *Phytopathology*, 1975, 65, 91-92.
[29] Porter, DM., Smith, DH. & Rodriguez-Kabana, R. (1982). Peanut Diseases. In HE; Pattee, & CT. Young (Eds.), *Peanut Science and Technology*, (326-410). Yoakum, TX: Am. Peanut Res. Educ. Soc., Inc.

[30] Report on the status of Arachis germplasm in the United States. http://www.ars-grin.gov/npgs/cgc_reports/Status11.pdf (accessed: 12/22/2009).

[31] Blankenship, PD; Cole, RJ; Sanders, TH; Hill, RA. Effect of geocarposphere temperature on pre-harvest colonization of drought-stressed peanuts by Aspergillus flavus and subsequent aflatoxin contamination. *Mycopathologia*, 1984, 85, 69-74.

[32] Lynch, RE; Wilson, DM. Enhanced infection of peanut, Arachis hypogaea L., seeds with Aspergillus flavus group fungi due to external scarification of peanut pods by the lesser cornstalk borer, Elasmopalpus lignosellus (Zeller). *Peanut Sci.*, 1991, 18, 110-116.

[33] Cole, RJ; Sanders, TH; Hill, RA; Blankenship, PD. Mean geocarposphere temperatures that induce preharvest aflatoxin contamination of peanuts under drought stress. *Mycopathologia*, 1985, 91, 41-46.

[34] Hammons, RO. Peanuts: Genetic vulnerability and breeding strategy. *Crop Sci.*, 1976, 16, 527-530.

[35] Knauft, DA; Gorbet, DW. Genetic diversity among peanut cultivars. *Crop Sci.*, 1989, 29, 1417-1422.

[36] Holbrook, CC; Stalker, HT. Peanut breeding and genetic resources. *Plant Breeding Rev.*, 2003, 22, 297-356.

[37] Sobolev, VS; Guo, BZ; Holbrook, CC; Lynch, RE. Interrelationship of phytoalexin production and disease resistance in selected peanut genotypes. *J. Agric. Food. Chem.*, 2007, 55, 2195-2200.

[38] Athar, M; Back, JH; Tang, X; Kim, KH; Kopelovich, L; Bickers, DR; Kim, AL. Resveratrol: A review of preclinical studies for human cancer prevention. *Toxicol. Appl. Pharmacol.*, 2007, 3, 274-283.

[39] de la Lastra, CA; Villegas, I. Resveratrol as an anti-inflammatory and anti-aging agent: mechanisms and clinical implications. *Mol. Nutr. Food Res.*, 2005, 49, 405-430.

[40] King, RE; Bomser, JA; Min, DB; King, RE; Bomser, JA; Min, DB. Bioactivity of resveratrol. *Comprehensive Rev. Food Sci. Food Safety*, 2006, 5, 65-70.

[41] Ulrich, S; Wolter, F; Stein, JM. Molecular mechanisms of the chemopreventive effects of resveratrol and its analogs in carcinogenesis. *Mol. Nutr. Food Res.*, 2005, 49, 452-461.

[42] Kadman-Zahavi, A. Persistence of Pfr and other manifestations of phytochrome in the nyctinastic leaflet movements of peanut (Arachis hypogaea L.). *Israel J. Bot.*, 1972, 21, 142-149.

[43] Ziv, MA; Halevy, H; Ashn, A. Phytohormones and light regulation of the growth habit in peanuts (Arachis hypogaea L.). *Plant and Cell Physiol.*, 1973, 14, 727-735.

[44] Ono, Y; Ozaki, K. Effects of shading treatment at early growth stage on growth and yield of peanut plants. *Proc. Crop Sci. Soc.* Japan, 1971, 40, 480-485.

[45] Hudgens, RE; McCloud, DE. The effect of low light intensity on flowering, yield and kernel size of Florunner peanut. *Proc. Soil Crop Sci. Soc. Fla.*, 1975 34, 176-178.

[46] Bell, MJ; Harch, G; Wright GC. Plant population studies on peanut (Arachis hypogaea L.) in subtropical Australia. 1. Growth under fully irrigated conditions. *Australian J. Exp. Agric.*, 1991, 31, 535-543.

[47] Bell, MJ; Wright, GC. Groundnut growth and development in contrasting environments. 1. Growth and plant density responses. *Expl. Agric.*, 1998, 34, 99-112.

[48] Karve, AD; Deshmukh, AK; Bhalerao, AC; Deshmukh, VA. Photomorphogenic regulation of reproductive development in groundnut and the significance of nyctinastic leaf movements. *New Phytol.*, 1984, 96, 535-543.

[49] Gautreau, J. Influence of climatic factors on the growth and development of early groundnut. *Oleagineux*, 1973, 28, 567-577.

[50] Ziv, M; Koller, D; Halevy, AH. Ethylene and the geotropic response of lateral branches in peanuts (Arachis hypogaea L.). *Plant and Cell Physiol.*, 1976, 17, 333-339.

[51] Chhabra, N; Malik, CP. Influence of spectral quality of light on pollen tube elongation in Arachis hypogaea. *Ann. Bot.*, 1978, 42, 1109-1117.

[52] Elkan, GH. (1995). Biological nitrogen fixation in peanuts. In HE; Pattee, HT. Stalker, (Eds.), *Advances in peanut science*, (286-300). Stillwater, OK. American Peanut Research and Education Society, Inc.

[53] Burris, RH. (1976). Nitrogen fixation. In J; Bonner, & JE. Varner, (Eds.), *Plant Biochemistry*, (887- 908). New York, NY: Academic Press, Inc.

[54] Bergersen, FJ. (1977). Physiological chemistry of dinitrogen fixation by legumes. In RWF; Hardy, WS. Silver, (Eds.), A Treatise on Dinitrogen Fixation: Section III. *Biology*, (519-555). New York, NY: John Wiley and Sons.

[55] Dadarwal, KR; Singh, CS; Subba Rao, NS. Modulation and serological studies of Rhizobia from six species of Arachis. *Plant Soil*, 1974, 40, 535-544.

[56] Weaver, RW. Effectiveness of rhizobia forming nodules on Texas grown peanuts. *Peanut Sci.*, 1974, 1, 23-25.

[57] Singh, CS; Dadarwal, KR; Subba Rao, NS. Effectiveness of rhizobia from wild species of Arachis on the cultivated species, A. hypogaea, and their physiological characteristics. *Zentralbl. Bakteriol. Parasitenkd. Infek. Hyg. Abt. II*, 1976, 131, 72-78.

[58] Caldwell, BE; Vest, HG. (1977). Genetic aspects of nodulation and dinitrogen fixation by legumes: The macrosymbiont. In RWF; Hardy, WS. Silver, (Eds.), A Treatise on Dinitrogen Fixation. Section III. *Biology*, (557-576). New York, NY: John Wiley and Sons.

[59] Wynne, JC; Elkan, GH; Schneeweis, TJ; Isleib, TG; Preston, CM; Meisner, CA. Increasing nitrogen fixation of the peanut. *Proc. Amer. Peanut Res. & Educ. Assoc.*, 1978, 10, 22-29.

[60] Hardy, RWF; Burns, RC; Hebert, RR; Holsten, RD; Jackson, EK. (1971). Biological nitrogen fixation; a key co world protein. In TA; Lie, EG. Mulder, (Eds.), Biological Nitrogen Fixation and Natural and Agricultural Habitats: *Plant and Soil*, (Special vol., 561-590).

[61] Hardy, RWF; Havelka, VD. (1976). Photosynthare as a major factor limiting nitrogen fixation by field-grown legumes with emphasis on soybean. In PN. Nutman, (Ed.), Symbiotic Nitrogen Fixation in Plants. (Intl. Biol. Prog. 7, 421-439). Cambridge, MA: Cambridge Univ. Press.

[62] Pate, JS. (1976). Physiology of the reaction of nodulated legumes to environment. In PN. Nutman, (Ed.), Symbiotic Nitrogen Fixation in Plants. (Intl. Biol. Prog. 7, 335-360). Cambridge, MA: Cambridge Univ. Press.

[63] Balandreau, JP; Miller, CR; Dommergues, YR. Diurnal variations of nitrogenase activity in the field. *Appl. Microbiol.*, 1974, 27, 662-665.

[64] Dart, P. (1977). Infection and development of leguminous nodules. In RWF; Hardy, WS. Sliver, (Eds.), A Treatise on Dinitrogen Fixation: Section III. *Biology*, (567-472). New York, NY: John Wiley and Sons.
[65] Gibson, AH. (1976). Recovery and compensation by nodulated legumes to environmental stress. In PN. Nutman, (Ed.), Symbiotic Nitrogen Fixation in Plants. (*Intl. Biol. Prog*., 7, 385-403). Cambridge, MA: Cambridge Univ. Press.
[66] Pate, JS. (1977). Functional biology of dinitrogen fixation by legumes. In RWF; Hardy, WS. Silver, (Eds.), A Treatise on Dinitrogen Fixation: Section III. *Biology*, (473-517). New York, NY: John Wiley and Sons.
[67] Jordan, DL. Chapter 3. *Peanut Production Practices*. 2008/10/23. Available from: URL: http://www.p2pays.org/ref/13/12671/chptr3.html
[68] Munger, P; Bleiholder, H; Hack, H; Hess, M; Stauss, R; van den Boom, T; Weber, E. Phenological growth stages of the peanut plant (Arachis hypogaea L.): Codification and description according to the BBCH scale, *J. Agron. Crop Sci*., 1998, 180, 101-107.
[69] Smith, BW. Arachis hypogaea. Aerial flower and subterranean fruit. *Amer. J. Bot*., 1950, 37, 802-815.
[70] Oakes, AJ. Pollen behavior in the peanut (Arachis hypogaea L.). *Agron. J*., 1958, 50, 387.
[71] Hammons, RO; Leuck, DB. Natural cross-pollination of the peanut, Arachis hypogaea L., in the presence of bees and thrips. *Agron. J*., 1966, 58, 396.
[72] Chapter 3: *Clover and Some Relatives*. 2008/10/23. Available from: URL: http://bee.airoot.com/content/pollination_handbook/peanut.html
[73] Rodger, JG; Balkwill, K; Gemmill, B. African Pollination Studies: where are the gaps? Int. *J. Tropical Insect Sci*., 2004, 24, 5-28.
[74] Srinivasalu, N; Chandrasekaran, NR. A note on natural crossing in groundnut, Arachis hypogaea Linn. *Sci. Cult*., 1958, 23, 650.
[75] Leuck, DB; Hammons, RO. Occurrence of typical flowers and some associated bees (Apoidea) in the peanut, Arachis hypogaea. *Agron. J*., 1969, 61, 958-960.
[76] Leuck, DB; Hammons, RO. Further evaluation of the role of bees in natural cross-pollination of the peanut, Arachis hypogaea L. *Agron. J*., 1965, 57, 94.
[77] Leuck, DB; Hammons, RO. Pollen-collecting activities of bees among peanut flowers. *J. Econ. Ent*., 1965, 58, 1028-1030.
[78] Girardeau, JH; Leuck, DB. Effect of mechanical and bee tripping on yield of the peanut. *J. Econ. Ent*., 1967, 60, 1454-1455.
[79] Sprenkel, RK. *Identification and monitoring of insect pests in peanut*. 2008/10/23. Available from: URL: http://edis.ifas.ufl.edu/IN176.
[80] Smith, JW; Barfield, C. (1982). Management of preharvest insects. In HE; Pattee, GT. Young, (Eds.), *Peanut Science and Technology*, (250-325). Yoakum, TX: American Peanut Research and Education Society, Inc.
[81] *Insect and mite control in field crops 2007 – readers' note*. 2008/10/23. Available from URL: http://www.dpi.nsw.gov.au/__data/assets/pdf_file/0008/173690/insect-mite-control-crops-pt4.pdf.
[82] Harborne, JB; Grayer, RJ. (1993). Flavonoids and insects. In J. B. Harborne (Ed.), The Flavonoids: *Advances in research since*, 1986. (678), London, UK: Chapman and Hall.
[83] Hermann, K. Flavonols and flavones in food plants: a review. *J Food Technol*., 1976, 11, 433-442.

[84] Havsteen, BH. The biochemistry and medical significance of the flavonoids. *Pharmacology & Therapeutics*, 2002, 96, 67-202.

[85] Harborne JB; Williams CA. Advances in flavonoid research since 1992. *Phytochemistry*, 2000, 55, 481-504.

[86] Werner, C; Hu, W; Lorenzi-Riatsch A; Hesse M. Di-coumaroylspermidines and tri-coumaroylspermidines in anthers of different species of the genus Aphelandra. *Phytochemistry*, 1995, 40, 461-465.

[87] Nimtz, M; Bokern, M; Meurer-Grimes, B. Minor hydroxycinnamic acid spermidines from pollen of Quercus dentate. *Phytochemistry*, 1996, 43, 487-489.

[88] Bokern, M; Witte, L; Wray, V; Nimtz, M; Meurer-Grimes, B. Trisubstituted hydroxycinnamic acid spermidines from Quercus dentata pollen. *Phytochemistry*, 1995, 39, 1371-1375.

[89] Zamble, A; Sahpaz, S; Hennebelle, T; Carato, P; Bailleul, F. N^1,N^5,N^{10}-Tris(4-hydroxycinnamoyl)spermidines from Microdesmis keayana Roots. *Chem. Biodiversity*, 2006, 3, 982-989.

[90] Strack, D; Eilert, U; Wray, V; Wolff, J; Jaggy, H. Tricoumaroylspermidine in flowers of Rosaceae. *Phytochemistry*, 1990, 29, 2893-2896.

[91] Meurer, B; Wray, V; Grotjahn, L; Wiermann, R; Strack, D. Hydroxycinnamic acid spermidine amides from pollen of Corylus avellana L. *Phytochemistry*, 1986, 25, 433-435.

[92] Meurer, B; Wiermann, R; Strack, D. Phenylpropanoid patterns in Fagales pollen and their phylogenetic relevance. *Phytochemistry*, 1988, 27, 823-828.

[93] Perez-Amador, MA; Carbonell, J; Navarro, JL; Moritz, T; Beale, MH; Lewis, MJ; Hedden, P. N^4-Hexanoylspermidine, a new polyamine-related compound that accumulates during ovary and petal senescence in pea. *Plant Physiol.*, 1996, 110, 1177-1186.

[94] Meurer, B; Wray, V; Wiermann, R; Strack, D. Hydroxycinnamic acid-spermidine amides from pollen of Alnus glutinosa, Betula verrucosa and Pterocarya fraxinifolia. *Phytochemistry*, 1988, 27, 839-843.

[95] Youhnovski, N; Filipov, S; Linden, A; Guggisberg, A; Werner, C; Hesse, M. Two macrocyclic spermine alkaloids from Aphelandra fuscopunctata (Acanthaceae). *Phytochemistry*, 1999, 52, 1717-1723.

[96] Martin, C; Kunesch, G; Martin-Tanguy, J; Negrel, J; Paynot, M; Carré, M. Effect of cinnamoyl putrescines on in vitro cell multiplication and differentiation of tobacco explants. *Plant Cell Rep.,* 1985, 4, 158-160.

[97] Smith, TA; Best, GR. Distribution of the hordatines in barley. *Phytochemistry*, 1978, 17, 1093-1098.

[98] Nomura, T; Sue, M; Horikoshi, R; Tebayashi, S; Ishihara, A; Endo, TR; Iwamura, H. Occurrence of hordatines, the barley antifungal compounds, in a wheat-barley chromosome addition line. *Genes Genet. Syst.*, 1999, 74, 99-103.

[99] Guillet, G; De Luca, V. Wound-inducible biosynthesis of phytoalexin hydroxycinnamic acid amides of tyramine in tryptophan and tyrosine decarboxylase transgenic tobacco lines. *Plant Physiol.*, 2005, 137, 692-699.

[100] Jin, S; Yoshida, M; Nakajima, T; Murai, A. Accumulation of hydroxycinnamic acid amides in winter wheat under snow. Biosci. *Biotechnol. Biochem.*, 2003, 67, 1245-1249.

[101] Facchini, PJ; Hagel, J; Zulak, KG. Hydroxycinnamic acid amide metabolism: physiology and biochemistry. *Canadian J. Bot.*, 2002, 80, 577-589.
[102] Walters, D; Meurer-Grimes, B; Rivora, I. Antifungal activity of three spermidine conjugates. *FEMS Microbiol. Lett.*, 2001, 201, 255-258.
[103] Klose, MK; Atkinson, JK; Mercier, AJ. Effects of a hydroxy-cinnamoyl conjugate of spermidine on arthropod neuromuscular junctions. *J. Comp. Physiol. A: Neuroethol. Sens. Neural Behav. Physiol.*, 2002, 187, 945-952.
[104] Lin, S; Mullin, CA. Lipid, polyamide, and flavonol phagostimulants for adult western corn rootworm from sunflower (Helianthus annuus L.) pollen. *J. Agric. Food Chem.*, 1999, 47, 1223-1229.
[105] Cabanne, F; Dalebroux, MA; Martin-Tanguy, J; Martin, C. Hydroxycinnamic acid amides and ripening to flower of Nicotiana tabacum var xanthi n.c. *Physiol. Plant.*, 1981, 53, 399-404.
[106] Martin-Tanguy, J; Perdrizet, E; Prevost, J; Martin, C. Hydroxycinnamic acid amides in fertile and cytoplasmic male sterile lines of maize. *Phytochemistry*, 1982, 21, 1939-1945.
[107] Ponchet, M; Martin-Tanguy, J; Marais, A; Martin, C. Hydroxycinnamoyl acid amides and aromatic amines in the inflorescences of some araceae species. *Phytochemistry*, 1980, 21, 2865-2869.
[108] Martin-Tanguy, J. Conjugated polyamines and reproductive development: biochemical, molecular and physiological approaches. *Physiol. Plant.*, 1997, 100, 675-688.
[109] Tarenghi, E; Martin-Tanguy, J. Polyamines, floral induction and floral development of strawberry (Fragaria ananassa Duch.). *Plant Growth Reg.*, 1995, 17, 157-165.
[110] Galston, A. W. & Flores, H. E. (1991). Polyamines and plant morphogenesis. In: RD; Slocum, & HE. Flores, (Eds), *Biochemistry and Physiology of Polyamines in Plants*, (175-187). Boca Raton, FL: CRC Press.
[111] Cabanne, F; Dalebroux, MA; Martin-Tanguy, J; Martin, C. Hydroxycinnamic acid amides and ripening of flower of Nicotiana tabacum var Xanthi n.c. *Physiol. Plant*, 1981, 53, 399-404.
[112] Martin-Tanguy, J. The occurrence and possible functions of hydroxycinnamic acid amides in plant. *Plant Growth Regul.*, 1985, 3, 381-399.
[113] Aribaud, M; Martin-Tanguy, J. Polyamine metabolism, floral initiation and floral development in Chrysanthemum (Chrysanthemum morijblium Ramat.). *Plant Growth Regul.*, 1994, 15, 23-51.
[114] Ma, C. M; Nakamura, N; Hattori, M. Inhibitory effects on HIV-1 protease of tri-p-coumaroylspermidine from Artemisia caruifolia and related amides. *Chem. Pharm. Bull.*, 2001, 49, 915-917.
[115] Funatogawa, K; Hayashi, S; Shimomura, H; Yoshida, T; Hatano, T; Ito, H; Hirai, Y. Antibacterial activity of hydrolyzable tannins derived from medicinal plants against Helicobacter pylori. *Microbiol. Immunol.*, 2004, 48, 251-261.
[116] Mullen, W; Boitier, A; Stewart, AJ; Crozier, A. Flavonoid metabolites in human plasma and urine after the consumption of red onions: analysis by liquid chromatography with photodiode array and full scan tandem mass spectrometric detection. *J. Chromatogr. A*, 2004, 1058, 163-168.

[117] Schieber, A; Keller, P; Streker, P; Klaiber, I; Carle, R. Detection of isorhamnetin glycosides in extracts of apples (Malus domestica cv. "Brettacher") by HPLC-PDA and HPLC-APCI-MS/MS. *Phytochem. Anal.*, 2002, 13, 87-94.

[118] Boumaza, O; Mekkiou, R; Seghiri, R; Sarri, D; Benayache, S; Garcia, VP; Bermejo, J; Benayache F. Flavonoids and isoflavonoids from Genista tricuspidata. *Chem. Nat. Comp.,* 2006, 42, 730-731.

[119] Bigler, L; Schnider, CF; Hu, W; Hesse, M. Electrospray-ionization mass spectrometry. Part 3. Acid-catalyzed isomerization of N,N'-bis[(E)-3-(4-hydroxyphenyl)prop-2-enoyl]spermidines by the Zip reaction. *Helv. Chim. Acta*, 1996, 79, 2152-2163.

[120] Hu, W; Reder, E; Hesse, M. Electrospray-ionization mass spectrometry. Part 2. Neighboring-group participation in the mass-spectral decomposition of 4-hydroxycinnamoyl-spermidines. *Helv. Chim. Acta*, 1996, 79, 2137-2151.

[121] The Wiley Registry of Mass Spectral Data (2000) (7th Edition). New York, NY: John Wiley and Sons.

[122] Lewis, FD; Elbert, JE; Upthagrove, AL; Hale, PD. Structure and photoisomerization of (E)- and (Z)-cinnamamides and their lewis acid complexes. *J. Org. Chem.*, 1991, 56, 553-561.

In: Organic Farming and Peanut Crops
Editors: D. C. Grossman, T. L. Barrios, pp. 103-122
ISBN: 978-1-60876-187-6

Chapter 3

FUNGAL AND MYCOTOXINS DISTRIBUTION IN PEANUT SAMPLES FROM SOWING TO HARVEST

Edlayne Gonçalez[1*], Joana D. Felicio[1], Maria Helena Rossi[1], Simone Aquino[2] and Benedito Corrêa[3]

[1]Centro de Pesquisa e Desenvolvimento de Sanidade Animal, Instituto Biológico, Av. Cons. Rodrigues Alves, 1252, São Paulo, Brazil, CEP 04014-002.

[2]Instituto Adolfo Lutz. Av. Dr. Arnaldo, 355, CEP 01246-902, Cerqueira César, São Paulo, Brazil.

[3]Instituto de Ciências Biomédicas, Av. Professor Lineu Prestes s/n, CEP 05508-900, Universidade de São Paulo, São Paulo, Brazil.

ABSTRACT

The elucidation of the contamination way of the toxigenic fungi and mycotoxins in peanuts focusing the environmental and field conditions is very important, so as to develop efficient control strategies. The peanuts "in natura" and your derived products are used in human and animal feed, besides the high frequency of toxigenic fungi like *Aspergillus flavus* and *Fusarium verticillioides* in Brazil and the importance of studies regarding mycotoxins motivated this work. Under these considerations, the present study aimed to: 1. Identify the mycoflora of the soil, air, flower, pegs, peanut kernels and hulls in the different maturation stages and after drying; 2. Determinate the occurrence of aflatoxins B_1, B_2, G_1 and G_2, cyclopiazonic acid (CPA) and fumonisins B_1 and B_2 in peanut kernels and hulls in different maturation stages and after drying.

The mycoflora study of the peanut kernels and hulls showed that *Fusarium* spp. and *Aspergillus flavus* were the most frequent fungi isolated. The presence of *A. flavus* also was detected in soil and air samples. *A. parasiticus* was isolated from the peanut kernels only after drying. The toxigenic potential tests of isolated strains demonstrated that 85.7 % of *A. flavus* analyzed showed the ability to produce aflatoxins, 38.0 % also produced CPA, 33.3 % produced both toxins and 4.7 % produced only CPA. It was observed that all *A. parasiticus* strains produced aflatoxins.

* Corresponding author. Telephone: +55-11-50871753; fax: +55-11-50871754,
E-mail address: goncalez@biologico.sp.gov.br

Aflatoxins and CPA were detected in 32 % of the peanut kernel samples in concentration that ranged from 4.20 μg/kg to 198.84 μg/kg and from 260 μg/kg to 600 μg/kg, respectively. These mycotoxins were detected from pod filling stage. Only aflatoxins were detected in the peanut hulls (24% of the samples) and the concentration ranged from 5.76 μg/kg to 218.52 μg/kg. In despite of two toxigenic *F. verticillioides* strains were isolated from peanut kernels and hulls, there were no fumonisins in these samples. Good agricultural practices are indicated to minimize the risks in this region and to avoid the hazards of mycotoxin contamination of peanut kernels, because the presence of toxigenic fungi *A. flavus* and *A. parasiticus* in soil.

Key words: Peanut, aflatoxin, cyclopiazonic acid, fumonisin, fungi

INTRODUCTION

Peanuts are an important food commodity well adapted in tropical and subtropical areas, and the principal exporting countries were the USA, India, Sudan, Senegal, Brazil, South Africa, Gambia, China and Egypt [1]. In Brazil, 83.2 % of the national production is located at São Paulo State and, as well in another country, the peanut kernels are used "in natura" or after industrial manufacture, resulting in oil and cake to animal feed [2, 3]. The peanut hulls are utilized as litter poultry and dairy cattle loafing, as fiber source to cattle, and their flour is used as one source of low caloric dietary fiber in some products for human, after controlling the presence of aflatoxins [4-6].

Thus, Peanuts are an important commodity, but very susceptible to fungal infestation and mycotoxins contamination. Mycotoxins are secondary metabolites of certain filamentous fungi that can be produced in foods as a result of fungal growth [7]. Since the recognition of hazards of mycotoxin contamination in food and feed commodities as a result of the discovery of aflatoxin in the early 1960's, some countries have conducted surveys on the incidence of mycotoxins in their agricultural produces [8]. The mycotoxigenic fungi involved in the human food chain belong mainly to the *Fusarium*, *Aspergillus*, *Penicillium* and *Alternaria* genera. The simultaneous presence of different mycotoxins in the same commodity produced by fungi belonging to different genera is not uncommon [9].

Examples of mycotoxins of great public health and agroeconomic significance include aflatoxin, ochratoxins, trichothecenes, zearalenone, fumonisins, tremorgenic toxins and ergot alkaloids [10].

The peanuts are commonly infected for *A. flavus* and *A. parasiticus*, as well as aflatoxins produced by these fungi, may occur before or after harvest. Besides aflatoxin, *A. flavus* produces also the cyclopiazonic acid (CPA), which is a toxic indole tetramic acid, that is naturally produced in agricultural commodities such as peanuts and corn, mostly as co-contaminant with aflatoxins [11]. Other mycotoxins were found in peanuts [12].

In Brazil, the presence of aflatoxins in peanuts and peanut products has been an alarming problem, although the extent and quantities found have been mostly low [13]. Environmental conditions and management practices during production, harvest, handling, and storage may affect the nature and degree of mycobiota contamination [14]. Soil serves as a major source of inocullum of the *A. flavus* and *A. parasiticus* in crops [15]. Mycotoxins may produce adverse effects on humans and animals, and since peanuts are an important food commodity. The carcinogenicity of the mycotoxins (the grater consequent implications for human health) evoked global concern [16].

The objectives of this study were: 1) to determine the mycloflora of soil, air, peanuts flowers, peg, seed and peanut kernels and hulls collected in the field at different maturity stages; 2) to determine the occurrence of aflatoxins B_1, B_2, G_1 and G_2 cyclopiazonic acid and fumonisins B_1 and B_2 in the seeds, at different stages of maturity of peanuts and the peanuts after drying.

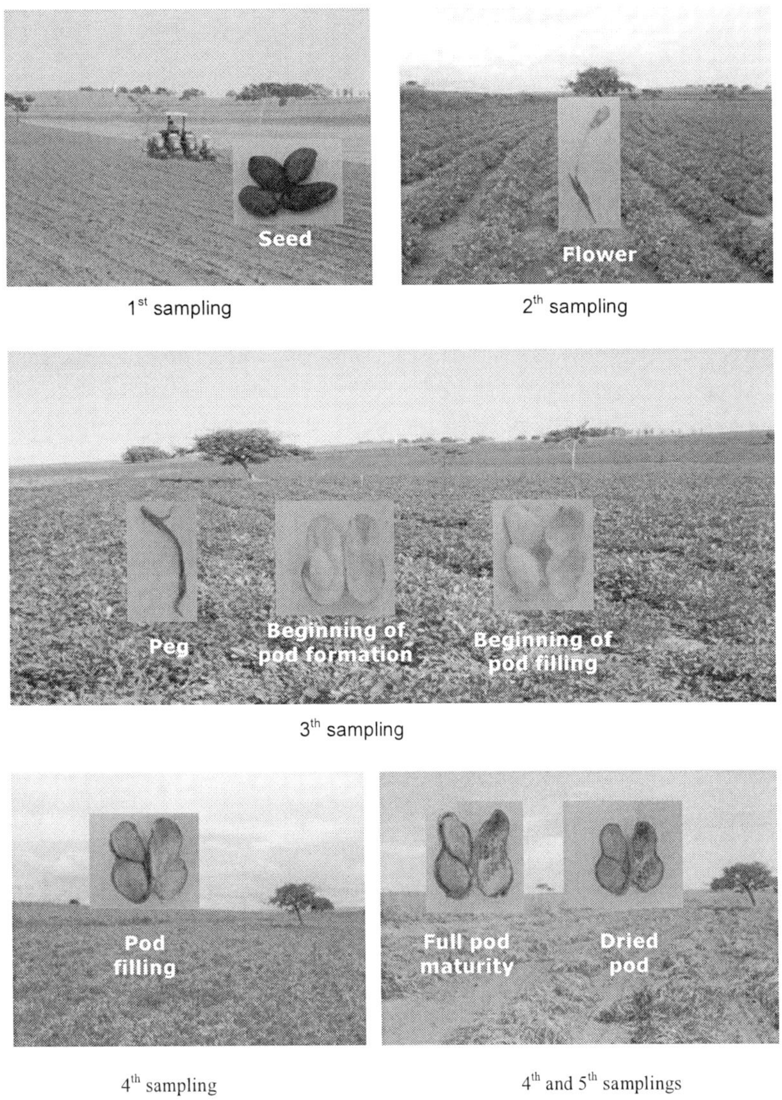

Figure 1. Peanuts Development Stage collected from Majeca farm.

Material and Methods

1. Peanuts Development Stage

The six samplings were based on peanuts phenological events (Figure 1). 1st) During sow planting, before germination (DS): for seed, air and soil; 2th) Beginning Bloom (BB): for flowers, air and soil; 3th) Peg (PG), at the Beginning of Pod formation (BP) and Beginning of Pod Filling (BPF): for pegs, pods, air and soil; 4th) Pod Filling (PF): for pods, air and soil; 5th) Full Pod Maturity (FPM): for pods, air and soil; 6th) Dried Pod (DP): for pods after drying.

2. Peanut Samples

The peanut seeds (Runner IAC 886) were cultivated in a farm located at Junqueirópolis, São Paulo State, Brazil, during the 2005/2006. The peanuts were sampled in four different maturity stages and after drying (Figure 1). The area was divided into 10 uniform parcels of 80 m^2, and five parcels were chosen for each sampling. Five nearly 1.0 kg samples were collected at the same time. In the all material collected, the shells of the pods were removed and the peanut kernel and hulls samples were assayed for mycobiota, aflatoxin, cyclopiazonic acid, fumonisin and water activity.

3. Soil Samples

Twenty-five samples were collected from the surface (0-5 cm depth), containing approximately 1.0 kg at each maturity stage. The bulk five soil samples were homogenized and assayed for mycobiota.

4. Water Activity Determinations

The water activity (a_w) of soil, seeds, flowers, peg, kernel and hull of peanut samples were determined by automatic analysis, using Aqualab CX-2 (Decagon Devices Inc., Pullman, WA).

5. Climatic Conditions of the Peanuts Field

The determination of the prevailing climatic factors was done with specific equipment such as thermometers and rain gauges. Soil temperatures were measured by thermometer, and the moisture (%) and pH were measured at 5 cm deep by a soil tester (Demetra E.M. System Soil Tester, Tokyo, Japan).

6. Mycobiota Analysis of Soil, Seed, Flowers, Peg and Peanuts Samples

6.1. Disinfection of seed, flowers, peg and pod of peanuts

From each sample, approximately 30 g subsamples of each vegetal material was disinfected by immersion in 0.4 % sodium hypochlorite solution for 3 minutes, followed by three rinses with sterile distilled water.

6.2. Fungal isolation from seed, flowers, peg and kernels and hulls of peanuts

Some 33 disinfected grains (11 kernels per dish) were sampled for mycobiota isolation. The same procedure was done for seed, flowers, pegs and hulls. Based on Pitt et al. [17], the first isolation was done on Dichloran Rose Bengal Chloramphenicol agar media (DRBC). Further selection of fungi producing aflatoxins was done on *A. flavus* and *A. parasiticus* Agar (AFPA). The fungal colonies recovered were identified according to recommended methods for each genus [18-20].

6.3. Isolation of mycobiota from soil samples

Soil samples were analyzed according to the method described by Swanson et al. [21] using the Martin medium for mycobiota isolation. The plates were incubated at 25°C for ten days, but the observations were made daily. The colonies were recovered and identified according to the recommended methods for each genus [18-20].

7. Determination of Wind-Dispersed Fungi

The samplings of wind-dispersed fungi were done before silk emergence. For each period, 10 Petri dishes containing sterile Sabouraud agar had the lid removed, were put at 1.50 cm above the ground lids were removed and the dishes kept open for 15 min to allow deposition of wind-dispersed fungal spores [22]. The plates were then incubated at 25°C for 7 days, and the fungal colonies identified according to the recommended methods for each genus [18-20]. The analysis of wind-dispersed fungi was carried out throughout the period of soil and peanut samplings.

8. Aflatoxin Analysis

The aflatoxins B_1, B_2, G_1 and G_2 in peanut kernel samples were extracted according to the method of AOAC 49.12.18 [23]. Briefly, 25 g of peanut kernels were added to 5 g NaCl and 125 mL of methanol/water solution (7:3 v/v) and stirred for 30 minutes. The extract was filtered through filter paper (Whatman 2V) and 30 mL of water were added to each 15 mL of the filtrate through glass microfiber filter paper (Whatman 943AH). Cleanup of the extracts was done with immunoaffinity columns (Aflatest, Vicam Somerville, MA, USA). Aliquots of 15 mL of each extract (equivalent to 1.0 g of the peanut kernels) were eluted in Aflatest cartridges at a flow rate of 2 drops/s. The cartridges were twice rinsed with 10 mL of water. The aflatoxins were eluted with 1.0 mL of HPLC grade methanol, at a flow rate of 1.0 drop/s.

This method is applicable for determination of aflatoxins B_1, B_2, G_1, and G_2 at ≥10 ng total aflatoxins/g in raw peanuts.

The same proceeding, with some modifications, was used to peanuts hulls. Briefly, 25 g of peanut hulls were ground with 10g NaCl, 250 mL of methanol/water solution (60:40 v/v) were added and the mixture was then stirred for 30 minutes. The extract was filter paper filtered and 20 mL of water were added to each 20 mL of the filtrate through glass microfiber paper. The cleanup of the extracts was done in the immunoaffinity column (Aflatest, Vicam Somerville, MA, USA), basically using aliquots of 4.0 mL of each extract which were eluted in Aflatest cartridges at a flow rate of 2 drops/s. The cartridges were rinsed with 5.0 mL of water, and the aflatoxins were eluted with 1.0 mL of methanol HPLC grade, at a flow rate of 1.0 drop/s. This method is applicable for determination of aflatoxins B_1, B_2, G_1, and G_2 at ≥2.0 ng total aflatoxins/g in peanut hulls.

8.1. HPLC conditions

A HPLC (Shimadzu model LC6AD) with UV detector (Shimadzu model SPD 6AV) and spectrofluorometric detector (Shimadzu model RF 551) and the Shimadzu class VP 5.02 program were used in this study. The mobile phase was methanol/acetonitrile/0.1% phosphoric acid (24:24:52 v/v/v) maintained at a flow rate of 1.0 mL/min [24]. The HPLC column was an ODS column, 250 mm x 4.6 mm (Shimadzu, Japan) kept at 30°C. A fluorescence detector set at 364 nm (excitation) and 440 nm (emission) was used to detect and quantify the aflatoxin. The calibration curves were set up for each type of aflatoxin by the external standard method with five standard concentrations: aflatoxin B_1: 8.09 ng/mL; 11.32 ng/mL; 16.18 ng/mL; 22.65 ng/mL and 32.36 ng/mL (r^2=0.997); aflatoxin B_2: 1.0 ng/mL; 1.4 ng/mL; 2.0 ng/mL; 2.8 ng/mL and 4.0 ng/mL (r^2= 0.999); aflatoxin G_1: 2.54 ng/mL; 3.55 ng/mL; 5.08 ng/mL; 7.11 ng/mL and 10.16 ng/mL (r^2= 0.996); aflatoxin G_2: 1.92 ng/mL; 2.67 ng/mL; 3.84 ng/mL; 5.34 ng/mL and 7.68 ng/mL (r^2= 0.999). The HPLC quantification and detection limits were: 3.5 ng/mL and 2.0 ng/mL for aflatoxin B_1; 0.1 ng/mL and 0.05 ng/mL for aflatoxin B_2; 2.0 ng/mL and 1.0 ng/mL for aflatoxin G_1 and 0.15 ng/mL and 0.05 ng/mL for aflatoxin G_2. Recoveries for standard aflatoxin B_1, B_2, G_1 and G_2 should be at least 90%, 80%, 90%, 65%, respectively, for both, kernel and hull samples.

8.2. Aflatoxin derivatisation

The presence of aflatoxins in the samples was confirmed according to the method of Tarín et al. [25]. An aliquot of 500 μL of each sample with aflatoxins was evaporated to dryness under nitrogen stream. The residue was dissolved in a mixture of 50 μL of trifluoroacetic acid and 200 μL of *n*-hexane at 40°C. The solution was shaken for 15 min and again dried under nitrogen stream and redissolved in 200 μL of acetronitrile/deionised water solution (30:70 v/v/v) for HPLC analysis.

9. Cyclopiazonic Acid (CPA) Analysis

The peanut kernels samples were analyzed for CPA basically according to the method of Urano et al. [25] with some modifications. A total of 25 g of peanut kernels were added to 100 mL of methanol 2% sodium bicarbonate solution (7:3 v/v), and stirred for 30 minutes. In the peanut hulls samples were used 200 mL. The extract was filtered through filter paper (Whatman 2V) and 50 mL of the kernels extracts and 100 mL for hulls extracts were transferred to a separatory funnel where 100 mL hexane were added, followed by gently shaking. The phases were allowed to separate and the hexane phase was discarded. 50 mL of 10% KCl solution were then added to the separatory funnel, and 2.0 mL of 6N HCl solution were added to lower the pH. CPA was then extracted with two 50 mL volumes of chloroform. The chloroform phases were collected, filtered through anhydrous sodium sulfate and evaporated under vacuum. The residue was dissolved in 10 mL of chloroform and passed through a Micotox® MS2300 cartridge (Micotox Ltda, Bogotá D.C., Colombia). The residues in the cartridges were washed sequentially with 10 mL of ethylic ether, 10 mL of chloroform/acetone (1:1v/v) and 10 mL of chloroform/methanol (95:5 v/v). Elute CPA-containing fraction with 10 mL of chloroform/methanol (75:25 v/v) at a flow rate of 2.0 mL/min, collecting eluate in 50 mL conical flask. Dry eluate under stream nitrogen at room temperature. The residue was dissolved with 1.0 mL of methanol HPLC grade. The limit of CPA determination was 50 ng/g [26].

9.1. HPLC conditions

The mobile phase consisted of acetonitrile/ammonium acetate buffer (0.05M, pH5) (8:2 v/v) [27] at a flow rate of 0.6 mL/min. A C8 column, 250mm x 4.6mm (Shimadzu, Japan) was used and the analyses were done at 284 nm. The calibration curve was established by the external standard method with five concentrations: 0.611 μg/mL; 1.223 μg/mL; 2.446 μg/mL; 4.892 μg/mL and 9.798 μg/mL (r^2=0.999). The HPLC quantification and detection limits for a standard CPA were, respectively, 0.03 μg/mL and 0.005 μg/mL. Recoveries for standard CPA were 70% and 62%, respectively, peanut kernel and hull samples.

10. Fumonisins Analysis

The samples of peanut kernels and hulls were analyzed for fumonisins B_1 and B_2 according to the method of Visconti et al.[28]. Briefly, 50 mL of acetonitrile/methanol/water (25:25:50 v/v/v) solution was added to each 20 g samples of peanut kernels and the mixture was shaken for 20 minutes in an orbital shaker. In the peanut hulls were used 10 g of the samples and 100 mL of acetonitrile/methanol/water (25:25:50 v/v/v) solution. The mixture was then centrifuged for 10 min at 2500 x g, and the supernatant filtered through filter paper (Whatman No. 4). The process was repeated again with the remaining solid material. The two filtrates were combined and 10 mL aliquots of the peanut kernels extract were mixed with 40 mL of phosphate buffered saline (PBS). For the peanut hulls extract a 20 mL aliquots were mixed with 20 mL of phosphate buffered saline (PBS) were used. The diluted extracts were microfiber filtered (Whatman No. 4) and 10 mL aliquots, for both (kernel and hull extracts), were submitted to cleanup by immunoaffinity column (Fumonitest, Vicam, Somerville, MA, USA) at a flow rate of 1-2 drops/s. The fumonisins were eluted with 1.5 mL of methanol HPLC grade, at a flow rate of 1 drop/s. The quantification limit was agreed Visconti et al. [28], at total levels from 0.5 μg/g to 2.0 μg/g.

10.1. HPLC conditions

Two hundred microliters of the final extract were derivatized with 50 µL of o-phthaldialdehyde (OPA) solution (40 mg of OPA dissolved in 1.0 mL of methanol and diluted in 5.0 mL of 0.1M sodium tetraborate containing 50 µL of mercaptoethanol). The product of this reaction was analyzed by a reverse phase isocratic HPLC system using a 150 x 4.6 mm C18 column (5ODS-20, Phenomenex) at a wavelength of 355 nm excitation and 440 nm emission. The mobile phase consisted of acetonitrile/water/acetic acid (520:480:5 v/v/v) solution at the 1.2 mL/min flow rate. The calibration curve was done by external standard method using four concentrations for each fumonisin: fumonisin B_1: 0.25 µg/mL; 0.5 µg/mL; 1.0 µg/mL and 2.0 µg/mL (r^2=0.996); fumonisin B_2: 0.125 µg/mL; 0.25 µg/mL; 0.5 µg/mL and 1.0 µg/mL (r^2=0.997). The HPLC quantification and detection limits were: 0.25 µg/mL and 0.025 µg/mL for fumonisin B_1 and 0.125 µg/mL and 0.0125 µg/mL for fumonisin B_2. Recoveries for standard fumonisin B_1 and B_2, should be at least 77% and 71%, respectively, for both, kernel and hull samples.

11. Evaluation of Toxigenic Potential

11.1 Aflatoxins

The fungi *A. flavus* (21 strains) and *A. parasiticus* (04 strains) were screened for the ability to produce aflatoxins by inoculation in coconut agar medium at pH 7.0 ±1.0. All plates were incubated at 25°C for 10 days and the average culture around it was transferred to glass flasks, weighted and macerated in chloroform at a ratio of 3.0 mL/g. The material was filtered and the filtrate was evaporated to dryness, and the residues were dissolved and made up to 1.0 mL in a volumetric flask with chloroform. Samples of 5.0 µL were spotted on silica gel-G thin layer plate which were developed using chloroform:acetone 9:1 (v/v) as solvent system. Concentrations of aflatoxins were determined by photodensitometry (photodensitometer Shimadzu, CS 9000), by comparing the area of the spots samples with standards.

11.2 Cyclopiazonic acid

Twenty one strains of *A. flavus* were screened for the ability to produce CPA by inoculation in PDA culture at 25°C for 5 days. The spores were harvested by adding 10 mL of sterile water containing 0.05% Tween 80, and 1.0 mL of the spores solution were inoculated in Czapeck-Dox medium (DIFCO). The cultures were incubated at 25°C for 12 days and the medium was filtrated and the CPA was extracted twice with 25 mL of chloroform. The extract was filtered and evaporated to dryness, and the residues were dissolved and made up to 1.0 mL of methanol HPLC grade. The CPA was quantified like described in the 9.1.

11.3 Fumonisins

Fusarium verticillioides (02 strains) were screened for the ability to produce fumonisin B_1 and B_2 by inoculation in PDA culture at 25°C for 5 days. The spores were harvested by adding 10 mL of sterile water containing 0.05% Tween 80, and 1.0 mL of the spores solution were inoculated in 50 g of sterilized rice and incubated at 25°C for 15 days and at 15°C for 15 days. The fumonisins were extracted of the rice with 100 mL of the acetonitrile:water solution 1:1 (v/v) and stirred for 30 minutes. The extract was filtered and 10 mL aliquot was submitted to cleanup by SAX cartridges (VARIAM). The residues in the cartridges were washed with 5.0 mL of the methanol:water solution 1:1 (v/v) and the fumonisins were eluated with 15 mL of methanol:acetic acid solution 99:1 (v/v). The extract was evaporated until dryness, and the residues were dissolved and made up to 1.0 mL of methanol HPLC grade. The fumonisins were quantified like described in the 10.1.

12. Statistical Analysis

Linear regression models were estimated to achieve the several **explanatory variables** and to account for the independent variables (*A. flavus* and *A. parasiticus* in peanuts kernel and hull, seed, flower, peg, soil and air; water activity; temperature, pH and moisture of soil; environmental temperature and rainfall index) on the dependent variables (aflatoxins and CPA in peanut kernels and hulls). The statistical analysis used the REG procedure of SAS® (Statistical Analysis Software).

RESULT AND DISCUSSION

The following mycobiota was isolated in the seed samples: non-sporulated fungi (36%), *Penicillium* spp. (15%), *Rhizopus* spp. (11%), *Fusarium* spp. (8%), *Phoma* spp. (7%), *Trichoderma* spp. (7%) *Cladosporium* spp. (4%), *A. flavus* (3%), *A. niger* (3%), *A. terreus* (2%) and *Fusarium verticillioides* (1%). Flowers and pegs showed 100% of *Fusarium* spp. The presence of *A. flavus* was detected in some seed samples (Runner IAC 886) and might have contaminated the soil and the peanut pods (Table 1).

Fungi isolated from the soil samples were *Penicillium* spp. (100% of samples), *Fusarium* spp. (84%), *A. terreus* (76%), *Mucor* spp. (28%), non-sporulated fungi (20%), *Cladosporium* spp. (16%), *A. flavus* (8%), *A. niger* (8%) e *Trichoderma* spp. (8%) (Table 1). The high frequency of *Penicillium* spp., *Fusarium* spp. and *Aspergillus* spp. in soil samples have already been described [29] and the high frequencies of *Penicillium* spp. has been reported also in São Paulo state [30]. *Aspergillus flavus* was isolated in the 1st and 4th samplings, and may be a result of contaminated seeds, or spores dispersed by the wind (Figure 2). Horn [31] reported that the soil is a primary source of inoculum for *A. flavus* and *A. parasiticus* in peanuts.

The occurrence of wind-dispersed fungi that were included *Cladosporium* spp. (58%), *Mucor* spp. (22%), *Aspergillus* spp. (18%), *Neurospora* spp. (16%), *Rhizopus* spp. (14%), *Helminthosporium* spp. (8%), *Penicillium* spp. (8%), non-sporulated fungi (8%), *Sepedonium* spp. (6%), *Alternaria* spp. (2%), *Epicoccum* spp. (2%) and *Trichoderma* spp. (2%) (Table 1). In Brazil, the presence of *Penicillium* spp. and *Aspergillus* spp. in air has previously been already reported by Almeida et al. [30].

In the peanut kernel samples the following mycobiota were isolated: *Fusarium* spp. (26%), *A. flavus* (17%), *A. terreus* (9.5%), *Penicillium* spp. (5%), *Rhizopus* spp. (4%), *A. parasiticus* (2.3%), *Curvularia* spp. (0.9%), *Beltrania* spp. (0.3%), *Mucor* spp. (0.3%), *A. niger* (0.3%) and *F. verticilliodes* (0.3%) (Table 1).

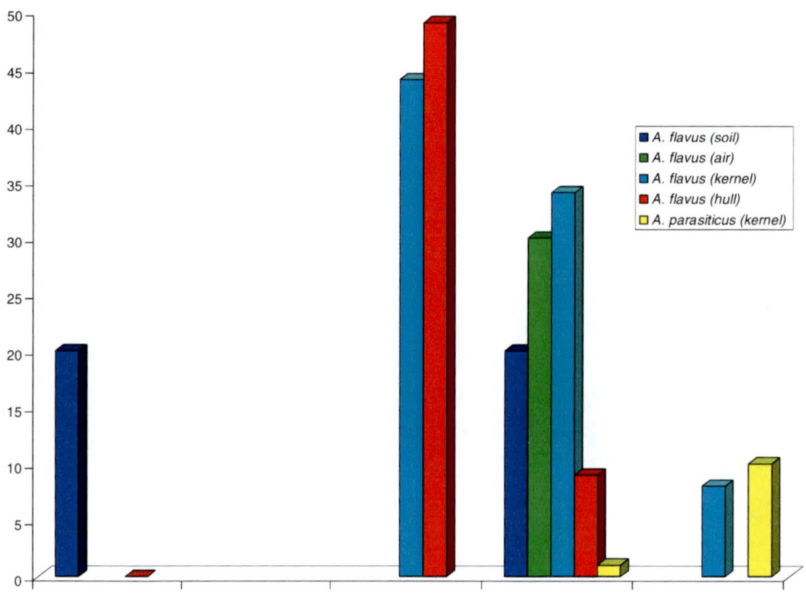

Figure 2. Relative frequency (%) of the *Aspergillus flavus* and *Aspergillus parasiticus* isolated from soil, air, kernel and hull samples at maturity stages (medium values, n=10 for air and n=5 for soil, kernel and hull).

The mycobiota isolated in the peanut hulls consisted in: *Fusarium* spp. (78.75%), *Rhizopus* spp. (14.10%), *A. flavus* (11.75%), *Absidia* spp. (8.00%), *A. niger* (2.50%), *Trichoderma* spp. (0.80%), *Penicillium* spp. (0.70%), *Scopulariopsis* spp. (0.15%) and *A. candidus* (0.13%) (Table 1).

The presence of *A. flavus* was the most important of fungal genera because of its known toxigenic potential and detection in the pod filling, full pod maturity and dried pod stages, and it was isolated only one sample of hull in the beginning of pod formation stage (1.2%) (Table 2 and Figure 2). Pitt et al. [17] have already reported that *A. flavus* was the dominant fungi isolated from peanut samples. *Aspergillus parasiticus* was the fungi most isolated in dry pod, detected in 80% of the kernels samples. The highest frequency of *A. flavus* in the peanut kernels and hulls was found in pod filling (43.9% and 49.0%), full pod maturity (33.6% and 8.8%) and dried pod stages (7.5% and 0%) (Table 2 and Figure 2). Moreover, *A. parasiticus* was also found more frequently than *A. flavus* at kernel samples, in dry pod, 10.3 % and 7.5% respectively, in spite of low water activity (0.71) (Table 2 and Figure 2). Isolation of aflatoxigenic fungi in low a_w (0.30 to 0.69) conditions has been reported previously [32], but in the hulls after the pod drying, when the a_w value was 0.61 (Table 2), *A. flavus* was not isolated. The average values of a_w in the hull samples, varied from 0.99 to 0.61 (Table 2), being 0.96 found in the pod filling and 0.90 in the full pod maturity stages and these values were important to determine the *A. flavus* contamination ($p<0.05$ e $r^2=0.62$).

Aspergillus flavus and *A. parasiticus* were not isolated in the stages of maturity with the highest water contents, i. e, a_w 0.98 (Table 2). The presence of *A. flavus* in hull samples in the stage forming of pod (with a_w 0.99) is explained by the contamination with soil residues on surface, once that the fungi were found in soil samples. The *A. flavus* was found only in the

1st sampling and *A. parasiticus* in the 1st and 4th samplings, and, as well as the presence in the seeds, could be responsible by the soil contamination (Tables 1 and 2 and Figure 2).

In the flowers and aerial pegs, *A. flavus* and *A. parasiticus* were not isolated (Table 1), indicating that the peanut fruits were not colonized by aflatoxigenic fungi by this way. Cole et al. [33] reported that aerial invasion of flowers; pegs or both were not a significant source of kernel infection.

The *A. flavus* presence in the 1st (2.0 x 10^3 cfu/g) and 4th (5.0 x 10^3 cfu/g) soil samplings should have contaminated the peanuts pods, as the soil is a primary reservoir for *A. flavus* and *A. parasiticus* and peanuts are an underground fruit, their pods should be directly exposed to the two soil species [15]. Densities of these fungi in soil vary greatly among fields and may influence the severity of peanut infection [34].

Table 1. Relative frequency (%) of fungal in seed, flowers, pods, soil, air, peanut kernels and hulls (medium values, n=5).

Fungi	Seed	Flower	Peg	Soil	Air	Kernel	Hull
Absidia spp.	-	-	-	-	-	-	8
Alternaria spp.	-	-	-	-	2	-	-
Aspergillus spp.	-	-	-	-	18	-	-
Beltrania spp.	-	-	-	-	-	0.3	-
Cladosporium spp.	4	-	-	16	58	-	-
Curvularia spp.	-	-	-	-	-	0.9	-
Epicoccum spp.	-	-	-	-	2	-	-
Fusarium spp.	8	100	100	84	-	26	78.75
Helminthosporium spp.	-	-	-	-	8	-	-
Mucor spp.	-	-	-	28	22	0.3	-
Neurospora spp.	-	-	-	-	16	-	-
Penicillium spp.	15	-	-	100	8	5	0.7
Phoma spp.	7	-	-	-	-	-	
Rhizopus spp.	11	-	-	-	14	4	14.10
Sepedonium spp.	-	-	-	-	6	-	-
Scopulariopsis spp.	-	-	-	-	-	-	0.15
Trichoderma spp.	7	-	-	8	2	-	0.8
A. candidus	-	-	-	-	-	-	0.13
A. flavus	3	-	-	8	-	17	11.75
A. niger	3	-	-	8	-	0.3	2.5
A. parasiticus	-	-	-	-	-	2.3	-
A. terreus	2	-	-	76	-	9.5	-
F. verticillioides	1	-	-	-	-	0.3	-
Non-sporulated fungi	36	-	-	20	8	-	-

The incidence of *Fusarium* spp. in all peanut stages of maturity was already related to higher a_w (0.98) and temperatures around 25°C [31]. As previously reported by Pitt et al. [17], *F. verticilloides* was occasionally isolated, but at very low levels (Table 1).

The mycotoxin analyses indicated that aflatoxins were present in 32% of the peanut kernel samples, including 4% in pod filling, 16% in full pod maturity and 12% in dried pod stages. In the peanut hull samples, 24% were contaminated with aflatoxins, including 4% in pod filling, 16% in full pod maturity and 4% in dried pod stages. No aflatoxins were detected in the beginning of pod formation and beginning of pod filling stages (Table 2 and Figure 3 and 4). Among the different aflatoxins, aflatoxin B_1 was present in all contaminated samples (kernels and hulls), at the highest amounts, from 3.79 μg/kg to 125.07 μg/kg by kernels and from 5.42 μg/kg to 218.53 μg/kg by hulls (Figure 3 and 4). Aflatoxin B_2 was also found in 100% of contaminated peanut kernel samples, but in concentrations from 0.07 μg/kg to 2.3 μg/kg (Figure 3), but in the hull samples, this toxin was only detected in pod filling and in full pod maturity stage (Figure 4). Aflatoxin G_1 and G_2 were detected, respectively, in 37.5% and 62.5% of the kernel contaminated samples. Their concentrations varied from detected, but out of quantification limit, to 66.25 μg/kg for AFG_1 and from 0.26 μg/kg to 5.19 μg/kg for AFG_2 (Figure 3). In the peanut hulls, aflatoxins G_1 was detected 50% of the samples, and the concentrations varied from detected, but out of quantification limit, to 48.46 μg/kg, and the aflatoxin G_2 was detected only in the full pod maturity stage (10.24 μg/kg) (Figure 4). In the full pod maturity hulls the water activity was also a very important factor for aflatoxins contamination ($p<0.05$ e $r^2=0.75$). Table 2 shows the amounts of the sum of aflatoxins detected in each maturation stage.

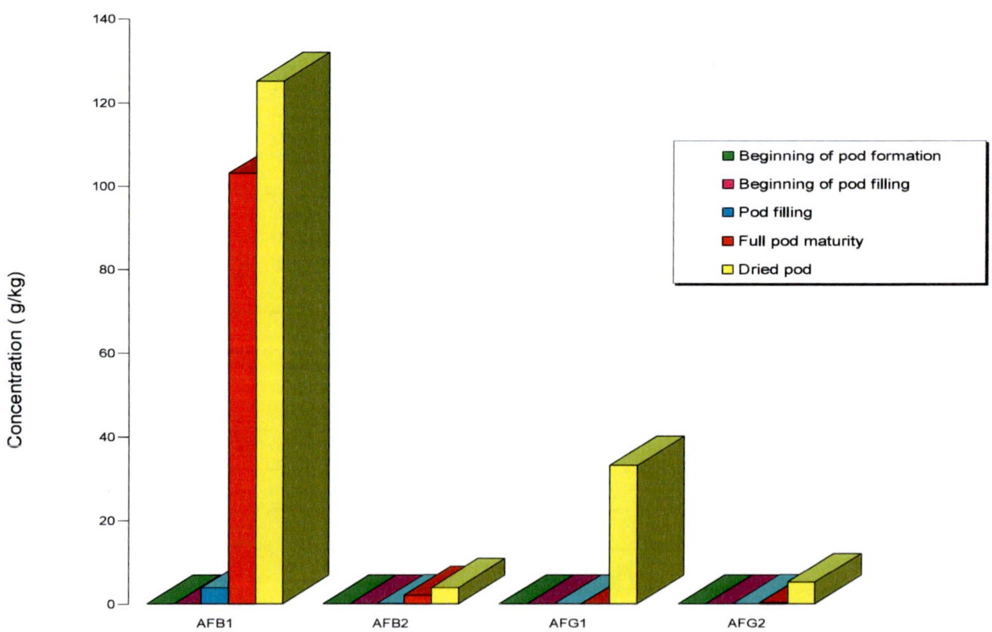

Figure 3. Aflatoxins B_1, B_2, G_1, G_2 concentration (μ g/kg) in peanut kernel samples at maturity stages (medium values, n=5).

Experiments using environmentally controlled plots have shown that *A. flavus* and *A. parasiticus* growth and aflatoxin formation in peanuts seeds are greatly enhanced by drought and elevated soil temperatures during the latter third of the growing season [35]. The something was observed in this study in the field situation (Table 2 and 3).

During the pod filling and full pod maturity stages of the peanuts maturation, when the rainfall in the field was low (from 0.43 to 5.5 mm), the kernels a_w were, respectively, 0.97 and 0.96 at pod filling and full pod maturity and 0.96 and 0.90 for hulls in the same maturation stages (Table 3), and the mean environmental and soil temperature were 27°C (Table 3). In the beginning of pod formation and beginning of pod filling stages, the mean environmental and soil temperatures were 25°C and 24°C, and the a_w of kernels was 0.98 and highest 0.97 for hulls (Table 2 and 3). In those periods, the rainfall index was 8.53mm (Table 3). Dorner et al. [35] reported that immature peanut pods are more resistant to fungi and aflatoxin contamination because they produce more total phytoalexins than mature peanuts, at high a_w. The authors also reported, that no phytoalexins production occurred when the a_w of kernels was below 0.95. Our results showed that the aflatoxin concentrations increased when the a_w of the kernels and hulls decreased (Table 2).

The highest aflatoxins concentration in the peanut hulls occurred at filling stage, followed to full pod maturity stage and after drying (Table 2 and Figure 4). The statistical analysis proved that the presence of *A. flavus* in the soil samples was an important variable for aflatoxins contamination in peanut hulls of full pod maturity stage ($p<0.05$ e $r^2=0.80$), and the kernels and hulls contamination were related ($p<0.05$ e $r^2=0.90$) in all maturity stages.

Sanders et al. [36] reported that the contamination of hulls by aflatoxins related to peanuts in the hull, which would be influenced by shelling efficiency as related to the quantity of peanuts, peanut parts and shriveled pods in the hulls.

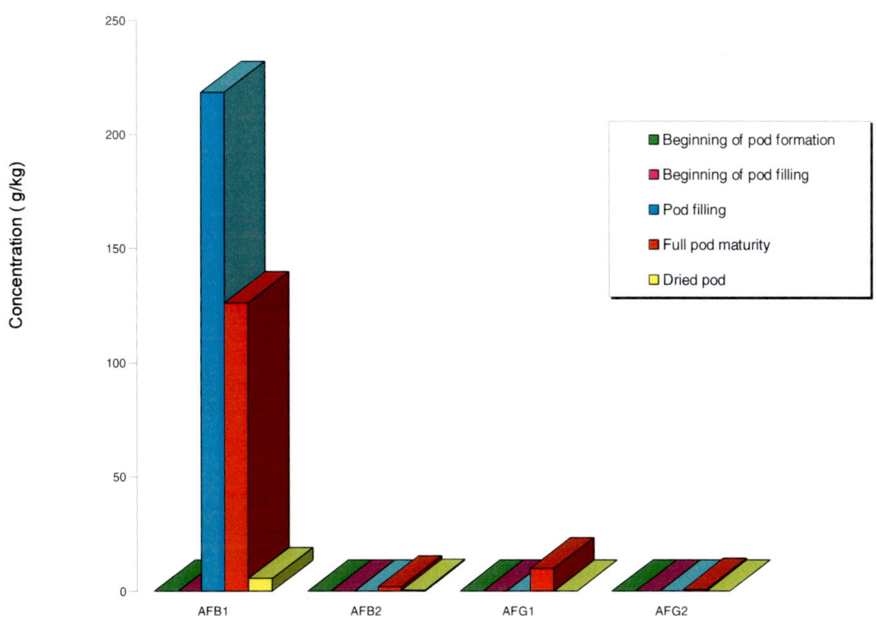

Figure 4. Aflatoxins B_1, B_2, G_1, G_2 concentration (μ g/kg) in peanut hull samples at maturity stages (medium values, n=5).

Severe drought stress on peanuts late in the growing season mostly resulted in decreased harvest, besides low quality yield, higher incidences of *A. flavus* group invasion and aflatoxin production [36]. Soil moisture and temperature have been shown as the major environmental factor involved in preharvest *A. flavus* and *A. parasiticus* invasion of peanuts and subsequent contamination with aflatoxin in extreme and prolonged drought stress [33].

The statistical analysis proved that a_w was an important variable to aflatoxin peanut kernels contamination at the full pod maturity stage ($p<0.05$ and $r^2=0.85$), but that the soil moisture was statically significant to kernel contamination by *A. flavus* in the pod filling and full pod maturity stages ($p<0.01$). Our results also agree with the literature because in peanuts with higher a_w (0.98), neither *A. flavus* or aflatoxins were detected in the kernels [31, 35]. On the other hand, the contamination by *A. flavus* and *A. parasiticus* showed positive correlation with aflatoxins in the dried peanut pod maturation stage ($p<0.01$ and $r^2=0.90$).

Like aflatoxins, CPA was detected in the different maturation stages. Thus, 32% of the samples were contaminated with CPA, with 8% in the filling pod stage, 16% in the full pod maturity and 8% in the dried pod. The concentrations of CPA varied from 206 µg/kg to 600 µg/kg (Table 2 and Figure 5). Maximum production was detected at a_w 0.96 and at 27°C of soil temperature, in the full pod maturity stage (Table 2 and 3). The statistical analysis showed the influence of the soil temperature in CPA production by *A. flavus* during this maturation stage ($r^2 = 0.48$).

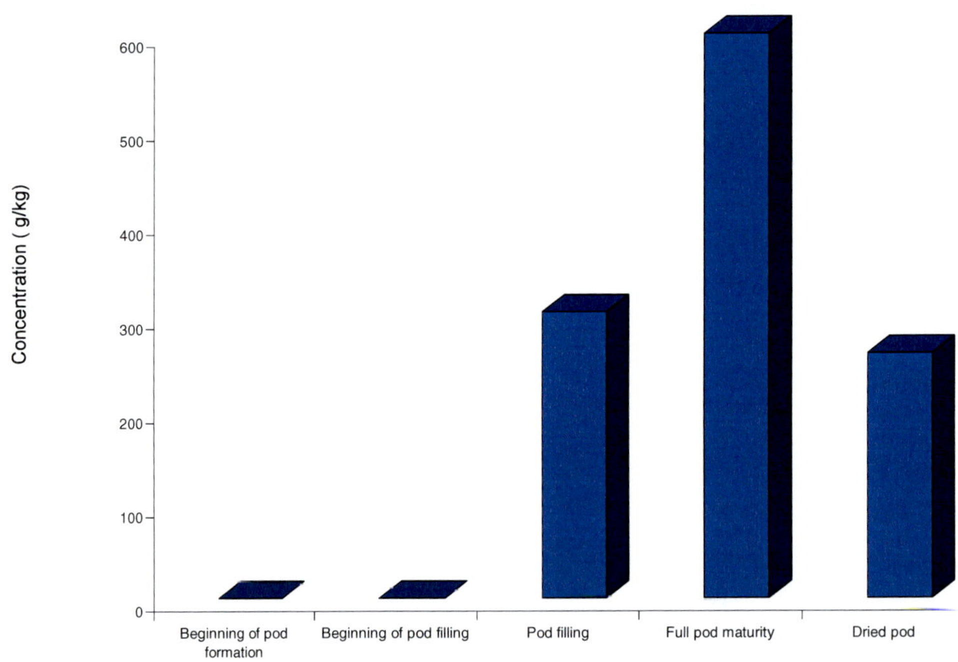

Figure 5. Cyclopiazonic acid (CPA) concentration (µ g/kg) in peanut kernel samples at maturity stages (medium values, n=5).

Table 2. Water activity (A_w), average of rainfall and environmental temperature, and concentrations of sum of aflatoxins and CPA. Relative frequency (%) of *A. flavus* and *A. parasiticus* and in peanuts kernel and hull are also showed (medium values, n=5).

Peanut Maturity Stage	A_w		Aflatoxins µg/kg (positive)		CPA µg/kg (positive)		*A. flavus* (%)		*A. parasiticus* (%)	
	kernel	hull	kernel	hull	kernel	hull	kernel	hull	kernel	hull
BP	0.98	0.99	ND*	ND*	ND*	ND*	--	1.2	--	--
BPF	0.98	0.97	ND*	ND*	ND*	ND*	--	--	--	--
PF	0.97	0.96	4.20 (1)	218.5 (1)	304 (2)	ND*	43.92	49.0	--	--
FPM	0.96	0.90	105.24 (4)	138.31 (4)	600 (4)	ND*	33.60	8.8	0.60	--
DP	0.71	0.61	198.84 (3)	5.77 (1)	260 (2)	ND*	7.55	--	10.30	--

* Not detected
BP: Beginning of Pod Formation
BPF Beginning of Pod Filling
PF: Pod Filling
FPM: Full Pod Maturity
DP: Dried Pod

Table 3. Soil data (temperature, moisture, pH and water activity) and relative frequency of *A. flavus*. Average of rainfall and environmental temperature (medium values, n=5).

Samplings	Soil temperature °C	Moisture (%)	pH	A_w	*A. flavus* %	Rainfall (mm)	Environmental Temperature °C
DS	26.4	15	6.9	0.99	20	8.53	25
BB	25.4	37.5	6.8	1.00	--	8.53	25
PG, BP, BPF	24.0	2.5	7.0	0.93	--	0.43	26.6
PF	27.2	5.6	7.0	0.99	20	5.5	26.5
FPM	27.0	3.7	7.0	0.98	--	0.0	26.5

A_w: Water activity
DS: During sow planting
BB: Beginning Bloom
PG: Peg
BP: Beginning of Pod formation
BPF: Beginning of Pod Filling
PF: Pod Filling
FPM: Full Pod Maturity

Natural co-occurrence of aflatoxins and CPA had been reported in peanuts [32, 37], but with different conditions for maximum production by *A. flavus* [38]. According to Vaamonde et al. [38] the presence of both toxins in food and feeds may result in additive or synergistic toxic effects. The optimum assigned temperatures for production of aflatoxins and CPA were 30°C and 25°C, respectively [39], which may explain the higher concentration of CPA than aflatoxins found in this study (Table 2) when the soil temperatures were ≤27°C (Table 3). Vaamonde et al. [38] reported that a_w 0.94 and 25°C as the optimal conditions for CPA production by *A. flavus*, after 28 days of peanut incubation in laboratory conditions and the authors concluded that the combination of high a_w and low temperature favoured high CPA production by *A. flavus*.

Table 4. Toxigenic potential of *Aspergillus flavus* strains isolated of seeds, soil, air, peanut kernel and hull samples. Concentration of aflatoxins B_1 and B_2 (µg/kg) and cyclopiazonic acid (µg/mL).

Material Proceed	Aflatoxin B_1 (µg/kg)	Aflatoxin B_2 (µg/kg)	Cyclopiazonic acid (µg/mL)
1st sampling (During sow planting)			
seed	6,700	260	11.85
soil	8.00	ND	ND
soil	ND	ND	ND
3th sampling (Beginning of Pod formation)			
hull	30	4.00	17.45
hull	190	1.00	ND
4th sampling (Pod Filling)			
soil	ND	ND	ND
air	20	ND	ND
kernel	ND	ND	ND
kernel	13,900	50	ND
kernel	ND	ND	4.48
Hull	7,700	200	2.36
hull	11,460	40	ND
5th sampling (Full Pod Maturity)			
kernel	18,560	150	8.07
kernel	4,300	220	ND
kernel	13,390	140	11.16
kernel	870	60	12.18
kernel	1,011	110	ND
hull	87	ND	ND
6th sampling (Dried Pod)			
kernel	600	10	ND
kernel	270	3.00	15.34
kernel	9,130	160	ND
hull	8,770	70	ND

ND: not detected

Fumonisins B_1 and B_2 were not detected independently of the maturity stage. The absence of fumonisins in peanut samples could be related to the low incidence *of F. verticillioides*, that was isolated in just one dried pod sample of the five analyzed. The a_w value found was 0.71 in this kernel sample, being below of minimum needed for this production toxin, between 0.92 to 0.96, according to Marín et al.[40], in spite of *F. verticillioides* strains isolated that showed the fumonisins production in the same conditions. The absence of fumonisin B_1 was also reported by Siame et al. [41], in peanuts from a Botswana market, however Sangare-Tigori et al., [12] were found 70% this toxin in peanut samples analyzed from Côte d'Ivoire.

Table 5. Toxigenic potential of *Aspergillus parasiticus* strains isolated of air and peanut kernel samples.

Material Proceed	AFB_1 (μg/kg)	AFB_2 (μg/kg)	AFG_1 (μg/kg)	AFG_2 (μg/kg)
1st sampling (During sow planting)				
air	24,340	250	300	45
4th sampling (Pod Filling)				
air	34,300	230	880	220
5th sampling (Full Pod Maturity)				
kernel	1,300	60	100	30
6th sampling (Dried Pod)				
kernel	5,060	30	1,210	900

The recoveries of CPA and fumonisins B_1 and B_2 were 62%, 84% and 71.35%, respectively. The results obtained showed that CPA or fumonisins were not detected in peanut hull samples, independently of the maturity stage.

The *A. flavus* strains analyzed from seed, soil, air, peanut kernels and hulls at different maturation stages, 85.7 % can be able to produce aflatoxins, 38.0 % produced CPA too, 33.3 % produced both toxins and 4.7% produced only CPA. The concentration varied from 8.0 μg/kg to 18,560 μg/kg and from 1.0 μg/kg to 260 μg/kg to AFB_1 and AFB_2, respectively (Table 4). The concentration of CPA varied from 2.3 μg/mL to 17.4 μg/mL (Table 4). The incidence of *A. flavus* strains produced of aflatoxins and CPA associated with peanut from field and storage, has also been shown by other researchers [38, 42].

All *A. parasiticus* strains produced aflatoxins, and the concentration of AFB_1, AFB_2, AFG_1 and AFG_2 varied from 1,300 μg/kg to 34,300 μg/kg, from 30 μg/kg to 250 μg/kg, from 100 μg/kg to 1,210 μg/kg and from 30 μg/kg to 900 μg/kg, respectively (Table 5).

The two strains of *F. verticillioides* were isolated from seed and kernel after dried and both producing fumonisin B_1 in the concentrations 320.41 μg/kg and 55,670 μg/kg, respectively. Only the strain from kernel produced FB_2 (5,023 μg/kg).

The maximum permissible aflatoxin level in Brazilian legislation for human feed is 20 μg/kg [43] and in our results 62.5% and 83.3% of the contaminated kernel and hulls samples, respectively, exceeded this value, indicating the need for constant monitoring of peanuts and peanut products, because this commodity is consumed in large amounts by the population and the hulls are used as poultry litter in the São Paulo regional chicken farms, so the aflatoxins continue to be a problem in Brazil. Although, CPA had been detected in food and there is no limit to CPA in Brazilian legislation.

Conclusion

Preharvest aflatoxin and CPA contamination of peanut is a great economic problem for the industry and a danger to human and animal health. Good management of the agricultural environment may be a way to reduce mycotoxins and toxigenic fungal contamination in peanuts at different maturity stages. The pods are exposed to different environmental conditions during their formation until harvest, and the optimal conditions for mycotoxin production and fungal growth are frequently found in the crop fields.

ACKNOWLEDGMENT

The authors are grateful to the Fundação de Amparo à Pesquisa do Estado de São Paulo (FAPESP) for supporting the research.

REFERENCES

[1] El-Maghraby, OMO; El-Maraghy, M. Mycobiota and mycotoxins of peanuts (Arachis hypogaea L.) seeds in Egypt. 1- sugar fungi and natural occurrence of mycotoxins. *Mycopathologia*, 1987, 98, 165-170.

[2] Fagundes, MH. Sementes de amendoim: alguns comentários [2006/08/28]. Avaible from: www.conab.gov.br.

[3] Santos, RC; Melo, PA; Brito, SFM; Moraes, JS. Fenologia de genótipos de amendoim dos tipos botânicos valência e Virgínia. *Pesquisa Agropecuaria Brasileira*, 1997, 32, 481-484.

[4] Collins, JL; Post, AR. Peanut hull flour as a potencial source of dietary fiber. *Journal of Food Science*, 1981, 46, 445-448.

[5] Lien, RJ; Hess, JB; Conner, DE; Wood, CW; Shelby, RA. Peanut hulls as a litter source for broiler breeder replacement pullets. *Poultry Science*, 1998, 77, 41-46.

[6] Hill, GM. Peanut by-products fed to cattle. *Veterinary Clinical of North America and Food Animal Practical*, 2002, 18, 295-315.

[7] Sweeney, M.J; Dobson, ADW. Mycotoxin production by Aspergillus, Fusarium and Penicillium species. *International Journal of Food Microbiology*, 1998, 43, 141-158.

[8] El-Maghragy, OMO; El-Maraghy, M. Mycoflora and mycotoxins of peanuts (Arachis hypogaea L.) seeds in Egypt. III- cellulose-decomposing and mycotoxins-producing fungi. *Mycopathologia*, 1988, 104, 19-24.

[9] Aresta, A; Cioffi, N; Palmisano, F; Zambonin, CG. Simultaneous determination of ochratoxin A and cyclopiazonic, mycophenolic, and tenuazonic acids in cornflakes by solid-phase microextraction coupled to high-performance liquid chromatography. *Journal Agricultural Food Chemistry*, 2003, 51, 5232-5237.

[10] Hussein, SH; Brasel, JM. Toxicity, metabolism, and impact of mycotoxins on humans and animals. *Toxicology*, 2001, 167, 101-134.

[11] Lisker, N; Michaeli, R; Frank, ZR. Mycotoxigenic potential of Aspergillus flavus strains isolated from groundnuts growing in Israel. *Mycopathologia*, 1993, 122, 177-183.

[12] Sangare-Tigori, B; Moukha, S; Kouadio, HJ; Betbeder, AM; Dano, DS; Creppy, EE. Co-occurrence of aflatoxin B1, fumonisin , ochratoxin A and zearalenone in cereals and peanuts from Côte d'Ivoire. *Food Additives and Contaminants*, 2006, 23, 1000-1007.

[13] Rodriguez-Amaya, DB; Sabino, M. Mycotoxins Research in Brazil: The Last Decade in Review. *Brazilian Journal of Microbiology*, 2002, 33, 1-11.

[14] Hill, RA; Blankenship, PD; Cole, RJ; Sanders, TH. Effects of soil moisture and temperature on preharvest invasion of peanuts by the Aspergillus flavus group and subsequent aflatoxin development. *Applied and Environmental Microbiology*, 1983, 45, 628-633.

[15] Horn, BW; Dorner, JW; Greene, RL; Blankenship, PD; Cole, RJ. Effect of Aspergillus parasiticus soil inocculum on invasion of peanut seeds. *Mycopathologia*, 1994, 125, 179-191.

[16] Mello, JPF; Macdonald, AMC. Mycotoxins. *Animal Feed Science Technology*, 1997, 69, 155-166.

[17] Pitt, JI; Hocking, AD; Bhudhasamai, K; Miscamble, BF; Wheeler, KA; Tanboon-Ek, P. The normal mycobiota of commodities from Thailand. 1. Nuts and oilseeds. *International Journal of Food Microbiology*, 1993, 20, 211-226.

[18] Barnett, HL & Hunter, BB. (1992). *Illustrated Genera of Imperfect Fungi*, Burguess, Minneapolis, MN.

[19] Raper, KB; Fennel, DI. *The genus Aspergillus*. Willians & Wilkins, Baltimore, MD. 1965.

[20] Nelson, PE; Touson, TA; *Marasas*, WFO. Fusarium species. An illustrated manual for identification. The Pennsylvania State University Press: London, U. K. 1983.

[21] Swanson, KM; Busta, FF; Petterson, EH; Johnson, MG. Colony Count Methods. In C; Vandezant, DS. Splittoesser, (Eds.), Compendium of methods for the microbiological examination of foods; New York: *American Public Health Association*, 1992.

[22] Gambale, W; Purchio, A; Paula, CR. Ação de fatores abióticos na dispersão aérea de fungos na cidade de São Paulo, Brasil. *Revista de Microbiologia*, 1983, 14, 204-214.

[23] AOAC. Official Method 991.3. Aflatoxins in corn, raw peanuts and peanut butter. Immunoaffinity column (aflatest) method. *AOAC International*, 2000, 49.12.18.

[24] Chan, D; Macdonald, SJ; Boughtflower, V; Brereton, P. Simultaneous determination of aflatoxins and ochratoxin A in food using a fully automated immunoaffinity column clean up and liquid chromatography-fluorescence detection. *Journal of Chromatography* A, 2004, 1059, 13-16.

[25] Tarin, A; Rosell, MG; Guardino, X. Use of high-performance liquid chromatography to assess airbone mycotoxins aflatoxins and ochratoxin A. *Journal of Chromatography A*, 2004, 1047, 235-240.

[26] Urano, T; Trucksess, MW; Beaver, RW; Wilson, DM; Dorener, JW; Dowell, FE. Cooccurrence of cyclopiazonic acid and aflatoxins in corn and peanuts. *Journal of AOAC International*, 1992, 75, 838-841.

[27] Losito, I; Monaci, L; Aresta, A; Zambonin, CG. LC-ion trap eletrospray MS-MS for determination of cyclopiazonic acid in milk samples. *Analyst*, 2002, 499-502.

[28] Visconti, A; Solfrizzo, M; Girolamo, A. Determination of fumonisins B1 and B2 in corn and corn flakes by liquid chromatography with immunoaffinity column cleanup: collaborative study. *Food Chemical Contaminants*, 2001, 84, 1828-1837.

[29] Klich, MA. Soil fungi of some low-altitude desert cotton fields and ability of their extracts to inhibit Aspergillus flavus. *Mycopathologia*, 1998, 142, 97-100.

[30] Almeida, AP; Fonseca, H; Fancelli, AL; Direito, GM; Ortega, EM; Corrêa, B. Mycobiota and fumonisin contamination in Brazilian corn from sowing to harvest. *Journal of Agricultural and Food Chemistry*, 2002, 50, 3877-3882.

[31] Horn, BW. Colonization of wounded peanuts seeds by soil fungi: selectivity for species from Aspergillus section Flavi. *Mycologia*, 2005, 97, 202-217.

[32] Mphane, FA; Siame, BA; Taylor, JE. Fungi, aflatoxin and cyclopiazonic acid Associated with peanut retailing in Botswana. *Journal of Food Protection*, 2004, 67, 96 102.

[33] Cole, RJ; Hill, RA; Blankenship, PD; Sanders, TH. Color mutants of Aspergillus flavus and Aspergillus parasiticus in a study of preharvest invasion of peanuts. *Applied and Environmental Microbiology*, 1986, 52, 1128-1131.

[34] Horn, BW. Relationship between soil densities of Aspergillus species and colonization of wounded peanut seeds. *Canadian Journal Microbioliogy*, 2006, 52, 951-960.

[35] Dorner, JW; Cole, RJ; Sanders, TH; Blankenship, PD. Interrelationship of kernel water activity, soil temperature, maturity, and phytoalexin production in preharvest aflatoxin contamination of drought-stressed peanuts. *Mycopathologia*, 1989, 105, 117-128.

[36] Sanders, TH; Blankenship, PD; Cole, RJ; Hill, RA. Effects of soil temperature and drought on peanuts pod and stem temperatures relative to Aspergillus flavus invasion and aflatoxin contamination. *Mycopathologia*, 1984, 86, 51-54.

[37] Fernández-Pinto, VE; Patriarca, A; Locani, O; Vaamonde, G. Natural co-occurrence of aflatoxin and cyclopiazonic acid in peanuts growth in Argentina. *Food Additives and Contaminants*, 2001, 18, 1017-1020.

[38] Vaamonde, G; Patriarca, A; Fernández-Pinto, VE. Effect of water activity and temperature on production of aflatoxin and cyclopiazonic acid by Aspergillus flavus in peanuts. *Advances in Experimental Medicine and Biology*, 2006, 571, 225-235.

[39] Gqaleni, N; Smith, JE; Lacey, J; Gettinby, G. Effects of temperature, water activity, and incubation time on production of aflatoxins and cyclopiazonic acid by isolate of Aspergillus flavus in surface agar culture. *Applied and Environmental Microbiology*, 1997, 63, 1048-1053.

[40] Marín, S; Sanchis, V; Vinas, I; Canela, R; Magan, N. Effect of water activity and temperature on growth and fumanisin B1 and B2 production by Fusarium moniliforme and Fusarium proliferatum on maize grain. *Letters in Applied Microbiology*, 1995, 21, 298-301.

[41] Siame, BA; Mpuchane, SF; Gashe, BA; Alltey, J; Teffera, G. Occurrence of aflatoxins, fumonisin B1 and zearalenone in foods and feeds in Botswana. *Journal of Food Protection*, 1998, 61, 1670-1673.

[42] Pildain, MB; Vaamonde, G; Cabral, D. Analysis of population structure of Aspergillus flavus peanut base don vegetative compatibility, geographic origin, mycotoxin and sclerotia production. *International Journal Food Microbiology*, 2004, 93, 31-40.

[43] Brasil – Leis e Decretos. Ministério da Agricultura. Portaria no. 183 de 21 de março de 1996, Diário Oficial da União de 25 de março, *Brasilia*, *Seção I*, 1996, 4929.

In: Organic Farming and Peanut Crops
Editors: D. C. Grossman, T. L. Barrios, pp. 123-136
ISBN: 978-1-60876-187-6

Chapter 4

HOW CAN THE TYPE OF DAIRY FARMING INFLUENCE THE BACTERIAL FLORA IN MILK?

A. Coorevits [1,2]***, V. De Jonghe*** [3]***, J. Vandroemme*** [4]***,***
A. Van Landschoot [1,2]***, M. Heyndrickx*** [3]***and P. De Vos***[2]

[1] Laboratory of Biochemistry and Brewing, Faculty of Applied Engineering Sciences, University College Ghent, member of Ghent University Association, Voskenslaan 270, 9000 Ghent, Belgium

[2] Laboratory of Microbiology, Department of Biochemistry and Microbiology, Ghent University, K.L. Ledeganckstraat 35, 9000 Ghent, Belgium

[3] Institute for Agricultural and Fisheries Research, Technology and Food Unit, Brusselsesteenweg 370, 9090 Melle, Belgium

[4] Institute for Agricultural and Fisheries Research, Plant Unit, Burgemeester Van Gansberghelaan 96/2, 9820 Merelbeke-Lemberge, Belgium

ABSTRACT

The dairy industry encounters important economical losses and health risks due to harmful spore-forming bacteria. Contamination occurs mainly at the milking stage and microbial numbers grow during conservation before pasteurisation or sterilisation. Some of the contaminants produce degrading enzymes such as lipases and proteases and render the milk unsuitable for human consumption. Others produce toxins and cause severe food illnesses and/or food poisoning. Spore-forming bacteria can survive heat treatments and jeopardize further production processes. Sources of the bacteria are diverse: feed and faeces are paramount, but air, grass, straw, milking equipment and the farmer's skin have been mentioned as well. Logically, the bacterial diversity of a contamination source will influence the resulting bacterial diversity of raw milk. Evidence suggests that differences in season, housing strategy and feed formulation contribute to the bacterial population of milk. Seeing that the operational management on dairy farms determines many of these factors, it seems obvious that this will also determine the bacterial diversity in the produced milk. To test this hypothesis, the diversity of aerobic endospore-forming bacteria in milk obtained from five conventional and five organic dairy farms in Flanders (Belgium) was compared. Although overall diversity was highly similar, some differences, possibly correlated to operational management, were observed. The greater number of thermotolerant bacteria in milk from conventional dairy farms for example, could be correlated with dietary differences. Cows from organic farms are fed with locally cultivated crops, while cows from conventional dairy farms consume commercially available feed concentrate, which is partially composed of pelleted (and hence heat-treated) and/or imported tropical ingredients. The observed higher occurrence of *Bacillus cereus* - a soil contaminant - in milk from organic dairy farms could be due to differences in housing strategy. Current evidence indicates that differences in operational management might influence the microbial diversity of raw milk, and as a consequence might have an impact on related economical losses and health risks. Further research in this field is needed and should include other bacterial contaminants and other countries.

Introduction

Organic farming is actually the oldest way of farming, i.e. without the use of chemicals (fertilizers) to enhance the production of meat, vegetables, milk and derivatives, and without the use of antibiotics and pesticides to decrease economical losses due to animal sickness and plant disease. Since the introduction of pesticides and fertilizers (mid 20th century), conventional agriculture established all over the world. However, despite the enormous economical advantages seen in the beginning of conventional farming, disadvantages such as soil pollution and insect pest resistance became apparent. This urged some environmentally aware farmers to go 'back to the roots' and prefer organic farming instead of conventional farming. For the dairy industry in particular, this implies animal's welfare and minimization of environmental impact being the main issues.

From the beginning of the re-introduction of organic farming, research has been focused on differences between this 'new' and the 'old' - conventional - type of dairy farming. Most papers report on differences in milk quality as outlined later on in this chapter. However, few research has dealt with the possible impact of these two types of dairy farming on the microbial aspect of dairy farming. Bacteria intrude raw milk via diverse contamination sources and can have a negative impact on milk quality and safety due to enzyme production causing milk spoilage on one hand and toxin production causing severe food illnesses on the other hand. These contaminants mainly are psychrotrophic pseudomonads or aerobic and anaerobic endospore-forming bacteria. To our knowledge, aerobic spore-formers are the only micro-organisms included in research comparing the microbial quality of milk from conventional and organic dairy farming thus far. Consequently, focus in this chapter will be on these aerobic spore-formers. Firstly, all problems the dairy industry encounters with this type of contaminants will be discussed. Secondly, differences between organic and conventional farming will be considered, with special attention to those differences possibly affecting the microbial flora of raw milk. In a third part, results on research conducted thus far will be disclosed. Finally, conclusions will be drawn on the microbial quality of milk from conventional and organic dairy farming.

Aerobic Endospore-Forming Bacteria in the Dairy Industry

Aerobic endospore-forming bacteria are the cause of many problems in the dairy industry concerning food safety and product quality. These problems arise due to the production of spoilage enzymes, toxins or through interference with cheese preparation due to gas formation or nitrate reduction.

Aerobic spore-formers contaminate raw milk through both the vegetative form and the heat resistant spore structure, and it is specifically this spore configuration that poses problems. A lot of studies have been performed on the contamination sources of these aerobic spore-formers (most research focussed on *Bacillus cereus*), but there is no consensus as to their origin in raw milk [64]. Very diverse contamination sources are indicated: soil [5, 8, 11, 29, 58], bedding [11, 13], feed [13, 29], hay [5], dust [5, 13], air [43], faeces [29, 58], dirty teats [11] or teat cups [2] and milking equipment [28, 29, 41]. Once these spores have intruded the milk, they cannot be destroyed by conventional heating processes - which are applied to destroy bacteria and thus assure safety of the product - such as pasteurization (e.g. 15 s 72°C) [1]. Some spores are even known to survive Ultra High Temperature (UHT)-treatment (e.g. 2-5 s 140-145°C) [44, 50, 51]. After the sterilisation step, spores can germinate into vegetative cells, a process which is even activated by heat; and subsequently grow in the milk or derived products, possibly causing problems mainly due to enzyme or toxin production.

It is well-known that aerobic spore-formers belonging to the genus *Bacillus* and related genera are involved in milk spoilage, hampering the quality of raw milk and derived products. Spoilage enzymes such as proteases, lipases and lecithinases are responsible for off-flavours and structural defects in pasteurized milk [37]. The most important spoilage organism in the dairy industry is undoubtedly *B. cereus*, causing 'bitty cream' (floating clumps of fat) due to lecithinase activity and 'sweet curdling' (curdling of the milk without acidification) due to protease activity [23]. The latter also causes bitter and rotten off-flavours, whereas fruity and rancid off-flavours are caused by lipolytic activity [37]. Furthermore, these enzymes have also been found in *Bacillus* species other than *B. cereus*, and appear to have heat resistant characteristics [7]. Another important trait of *B. cereus,* and some other bacilli too, is their ability to grow at the storage temperature of milk (4-7°C), which mainly determines the shelf life of pasteurised milk and derived milk products [37, 58].

Spoilage caused by aerobic spore-formers is mostly limited to pasteurized milk. Spoilage of UHT and sterilized milk only occurs occasionally and can mostly be attributed to recontamination with proteolytic *Bacillus* species or Gram-negative spoilage organisms during the filling step [14, 52, 66]. Despite this fact, there are problems with aerobic spore-formers in UHT-milk, not necessarily related to spoilage but more to a non-achievement of sterility standards. EG-regulation 92/46 requires that the number of colonies counted from incubated (30°C during 15 days) unopened UHT-cartons, does not exceed 10 colony forming units (cfu) per 0,1 mL. This requirement could not be met in Italy, Austria (1985) and Germany (1990) and later-on in other European (France, Benelux, Spain) and non-European (Mexico, USA) countries [20, 50]. Hammer *et al.* [20] linked this problem to highly heat resistant (HHR) spores of a mesophilic micro-organism described later as *Bacillus sporothermodurans* [44]. Since this organism does not have any pathogenic or toxic activity

[20, 21], and also only rarely causes minor spoilage defects in the form of a slight pink discoloration [25, 32], this is not so much an issue from a hygienic point of view, but all the more from a technological perspective, being a mere sterility issue.

Another issue regarding aerobic spore-formers in milk is the possible production of toxins. *Bacillus cereus* is a well known food pathogen that can cause two types of food poisoning syndromes: an emetic and a diarrhoeal type. The former is caused by a small ring-formed heat-stable dodecadepsipeptide named cereulide. The diarrhoeal syndrome is caused by vegetative cells that are thought to produce a variety of heat-labile enterotoxins in the small intestine [18]. Three pore-forming cytotoxins have been associated with diarrhoeal disease: haemolysin BL (Hbl) and nonhaemolytic enterotoxin (Nhe), which are both two homologous three-component toxins, and cytotoxin K (CytK), a single component cytotoxin. Two different forms for the latter have been described, the highly cytotoxic CytK-1 and the moderate cytotoxic CytK-2 variant, encoded by *cytK-1* and *cytK-2* genes [13]. The cytK-1 variant has thus far only been detected in a limited number of *B. cereus* strains, that have been proposed to form a novel bacterial species, for which the name "*Bacillus cytotoxis*" or "*Bacillus cytotoxicus*" is suggested [31].

Although *Bacillus* species other than *B. cereus* have been incriminated as food poisoning agents, the link between toxin production and foodborn illness has not been fully established. Increasing evidence for the production of both heat-stable and heat-labile toxins is becoming apparent through cellular assays that confirm both production and functionality of the toxins. In these assays, strains identified as *Brevibacillus brevis*, *Bacillus circulans*, *Bacillus lentus*, *Bacillus subtilis* [3], *Bacillus licheniformis* [3, 33], *Bacillus pumilus* [33] and *Bacillus amyloliquefaciens* [45] are shown to produce heat-labile enterotoxins. Strains identified as *B. licheniformis* [3, 17, 38, 49, 57], *B. pumilus* [16, 17, 55], *Bacillus mojavensis* [15], *B. subtilis* [3, 17], *Bacillus simplex* [57], *Bacillus firmus* [57], *Bacillus megaterium* [57], *B. brevis* [3], *B. circulans* [3] and *B. lentus* [3] can produce novel heat-stable toxins, resembling the physico-chemical characteristics of cereulide [17, 49, 55, 57]. Some of these toxins, heat-labile as well as heat-stabile toxins, have been characterised as surfactin isoforms, namely lichenysin (described in *B. licheniformis* [38, 49]) and pumilacidin (described in *B. pumilus* [55]). The surfactin superfamily is a family of structurally diverse, low molecular weight cyclic lactonic lipopeptides, that is well-described in strains of the *B. subtilis* group.

Another concern regarding aerobic spore-forming bacteria is interference with cheese production through nitrate reduction, whether or not in combination with gas production during fermentation. Nitrate is often added to raw milk during the cheese preparation process to prevent growth and germination of anaerobic spore-forming species belonging to the genus *Clostridium*. Specifically *Clostridium tyrobutyicum* can grow fermentatively with gas production, thereby causing 'late blowing' defects in Gouda type cheeses [26]. The reduction of nitrate to nitrite or nitrogenous gasses, is a process carried out by many *Bacillus* species [60], thus nullifying the growth inhibitory effect of supplemented nitrate. Recently, it became apparent that also gas producing aerobic spore-formers, namely *Paenibacillus polymyxa* and *Paenibacillus macerans*, can be directly implicated in this type of spoilage in Argentinian Cremoso and Mozarella cheeses [47].

Even though the dairy industry is often confronted with severe implications caused by aerobic spore-forming bacteria, their identity is most often either not determined onto the species level, or not reliably determined because identification has been based on phenotypic and biochemical characteristics of the isolates only [36]. Based on these inadequate identifications, the spoilage potential was thought to be restricted to *B. circulans*, *Bacillus coagulans*, *Brevibacillus laterosporus* [53], *P. polymyxa* [60] and *B. cereus* [42]. In view of current taxonomic insights in the aerobic spore-forming microbiota, up to 48% of all aerobic spore-forming isolates from raw milk could not be identified in some previous studies [19, 56]. It is clear that future diversity studies should focus on these yet unknown groups of bacteria, enabling an important improvement of our knowledge on contaminating aerobic spore-formers in the dairy industry.

The previous paragraphs indicate that problems with aerobic spore-formers encountered by the dairy industry might be affected by farm management (e.g. various types of feed). Indeed, bacteria intrude raw milk via various contamination sources as feed, hay, bedding, grass, and it can be expected that differences in these contamination sources between individual farms or between farm types might consequently have an impact on the composition of the microbiota of raw milk, thus affecting the quality and safety of milk and derived products [9]. One way to test this hypothesis could be by comparing the microbial load and diversity of milk from two types of farm management (i.e. organic versus conventional dairy farming) which may show differences in the relative importance of the above mentioned contamination sources.

Organic versus Conventional Dairy Farming

Organic farming aims at sustainable farming. It emphasises the animal's welfare and the use of local natural resources as much as possible, excluding chemical fertilizers and pesticides. The main principle of this type of farming is to minimize negative effects on the environment, preserve natural resources and maintain biological diversity on farms and their neighbourhood. Organic farming is gaining popularity over the last decade, however, in 2005, only 0.5% of all farming was regarded as organic in the U.S. and around 4% in the E.U. [34, 62]. This rather low percentage may be due to high managerial costs, risks of shifting to a new way of farming, limited awareness of organic farming systems and lack of marketing and infrastructure [62]. To obtain the label of organic, farms are obligated to follow some rules (determined by the Council Regulation in Europe and by the United States Department of Agriculture in America), which are created to preserve and protect the organic farming method. These rules imply, specifically for the dairy sector, restrictions on feed and housing strategy of animals, and also on health strategy.

Regarding the feed, organic farmers are compelled to use mainly locally cultivated crops, rather than imported feed concentrate. Even more, since 2007, all feed offered on organic dairy farms must be 100% organic in origin [34], precluding the use of chemical fertilizers and pesticides. The diet of animals from organic farms thus mainly consists of grass, cereals, fresh vegetables, barley and trefoil. On the other hand, cattle from conventional dairy farms are fed conventional crops, usually grain, supplemented with vitamins, minerals and feed concentrate. The latter provides the animals with the necessary proteins and is often composed of imported soy or manioc and tropical waste ingredients from products such as coconut meal and citrus pulp.

The regulations also specify a minimum space allocation per animal. Cattle from organic farms usually has more space, with larger stables and access to the outdoors. These measurements should have a positive impact on the animals' welfare, and indeed it has already been shown that reducing stocking density - which is a logical consequence of providing more space per head - does have a positive impact on dairy cow welfare [27]. High stocking densities however, affect a cow's ability to access feeders and lying areas [30].

Regarding health strategy, disease prevention rather than cure is emphasised in organic farming. Animals labelled as organic may not receive hormones nor antibiotics. Off course, organic farmers are allowed (or better obliged) to treat sick or injured animals with a prohibited medication, but these animals are not considered organic anymore. This lower treatment policy of organic herds might raise the concern that udder health is worse compared to conventional cattle. However, comparative studies on somatic cell counts of animals, which are used as an indicator of mastitis, reveal no significant difference between organic and conventional herds [22, 67]. In addition, the occurrence of mastitis, ketosis and milk fever was less abundant in organic herds, indicating an overall better condition of the cows compared to cows of conventional farms [22]. This was confirmed by a Danish study on herd health, stating fewer treatments for retained placenta and ketosis in organic herds [4]. In contradiction, Hörning *et al.* [24] reported that cows on German organic farms seem to be more susceptible to mastitis, indicated by higher somatic cell counts, but fewer fertility problems, and a longer life span. This discrepancy probably could be due to an underestimation of the occurrence of mastitis in organic herds. After all, the use of alternative herbal medicine and homeopathic treatments for mastitis - mostly applied in organic herds - is not recorded in databases, while antibiotic treatment - mostly applied in conventional herds - is always recorded [61].

Some studies report on the adoption of these 'rules for organic farming' into practice by interview questionnaire with the farmers. Zwald *et al.* [67] reported on the following differences between organic and conventional dairy farms in the US: organic herds are usually smaller and have a lower milk production, use of antibiotics is reduced, and is never used to treat mastitis in organic farms, calves are placed separately from the organic cows with a separate water supply which reduces the risk of disease transfer and conventional farms tend to buy their feed and concentrate, rather than producing it themselves. Coorevits *et al.* [9] described differences in feeding and housing strategy between organic and conventional dairy farming in Flanders, Belgium. Organic farmers indeed feed their cows locally cultivated crops such as maize, trefoil, barley, wheat, grass, carrots, turnip and chicory, supplemented with minerals, salt and seaweed calcium, while conventional farmers feed their cows with maize, grass and wheat supplemented with feed concentrate containing components including or derived from soy, rape, coconut, citrus and cacao. Differences in housing strategy were most obvious during summer season. During winter season, cows were mostly stabled day and night on both type of farms. In late summer/autumn, cows from organic farms more often grazed outside. Langford *et al.* [30] found that main differences between organic and conventional farming in the UK were milk yield (which was lower in organic herds), level and composition of used concentrate feed, management of heifers and calving, and use of alternative treatments to prevent and treat mastitis.

It can be concluded that management practices on organic dairy farms are in compliance with the regulations enforced to protect and preserve organic farming, and differ from management practices on conventional dairy farms, which is most obvious when comparing animal disease treatment, feed and housing strategy.

Various studies have dealt with the influence of these two different types of management on the animal's welfare, and on the quality of milk. Butler *et al.* [6] for instance, showed an influence of the length of the grazing period, and diet composition on sensory and nutritional quality of milk. Vicini *et al.* [65] reported on a greater protein concentration in milk from organic farms, which could be due to the use of another breed, or to the lower milk production by organic herds, implying a higher protein level. However, in the Netherlands, Nauta *et al.* [40] mentioned a highly significant decrease in protein percentage and also in fat content due to conversion from conventional to organic dairy farming. Very recently, Molkentin [39] clearly showed a difference in fatty acid composition of milk from organic and conventional dairy farms (e.g. higher levels of α-linolenic acid), which might be used to discriminate between the two types of milk. To our knowledge however, almost no studies report on differences in microbial quality of different types of milk. Only Vicini *et al.* [65] mention a non-significant lower bacterial count in milk from conventional farms, but no efforts to further characterize this bacterial flora have been made.

The hypothesis of our recent study [9] was that more detailed information (identification and characteristics) about these aerobic spore-formers would allow to trace the possible contamination sources, and find a relationship between these sources and the management practices on conventional and organic dairy farms. After all, a better insight in the factors influencing the introduction of spores into raw milk is crucial to take necessary actions to improve the quality and safety of milk.

Various papers report on correlations between micro-organisms found in raw milk and their contamination sources. For instance, soil has been identified as a major source of contamination of raw milk with *B. cereus* spores, the contamination being even more severe when cows grazed outside [8, 46]. Seasonal variations in psychrotolerant spore content have been reported in several investigations [5, 19, 46, 58], indicating a possible effect of pasturing, and thus grass and soil, as a contamination source [54]. As cows are more often grazing outside during summer period, the contamination sources soil and pasture can be linked to seasonal variation, and also to the housing strategy of the animals. Indeed, cows from organic farms are more often grazing outside than cows from conventional farms.

The presence of spore-formers in feed, in particular in silage and food concentrate, has been shown by Slaghuis *et al.* [54] and Vaerewijck *et al.* [63], while the correlation between spores in silage and spores in raw milk has been demonstrated by te Giffel *et al.* [59]. This can also be linked to management practice differences between organic and conventional dairy farming as conventional farmers are allowed to use commercially available feed concentrate, while the diet of organic cows is restricted to locally cultivated crops, without food additives. It thus seems obvious that operational management on dairy farms, which is related to contamination sources, will influence the aerobic spore-forming microbiota of raw milk. To investigate this hypothesis, the aerobic spore-forming flora of milk samples obtained from organic and conventional dairy farms was analysed [9].

Differences in Aerobic Spore-Forming Flora Between Organic and Conventional Raw Milk Samples

Before elaborating the results of that study, a short outline of the approach used to isolate and identify the spore-formers in organic and conventional raw milk samples is given, as already described [9].

Five milk samples from each type of dairy farm were collected in the region of Flanders, Belgium, and analysed during a summer and a winter campaign. All samples were heat-treated (10 min at 80°C) to select for spore-formers in particular, and were streaked in parallel on three different media [Brain Heart Infusion Agar (BHI), Nutrient Agar supplemented with egg yolk (NAE) and Skim Milk Powder Agar (SM)], with the latter two being differential media indicating lecithinase (NAE) and protease (SM) activity of isolates, and incubated at three different temperatures to select for psychrotrophic (20°C), mesophilic (37°C) and thermophilic (55°C) bacteria. Bacterial isolates were identified polyphasically. At first, fatty acid profiles of all isolates were generated to obtain a preliminary identification into more manageable groups of isolates. Subsequently, partial 16S rRNA gene sequences of representative isolates of each group were generated. A species allocation was obtained by using the online FASTA tool of EMBL (http://www.ebi.ac.uk/fasta33/), and comparison of the query sequence with 16S rRNA gene sequences of type strains of valid species. Furthermore, all isolates obtained after incubation at 20°C were screened for their potential to grow at refrigeration temperatures, the storage temperature of pasteurised milk (7 days at 6 ± 1°C).

Overall, the isolates turned out to belong to species such as *B. licheniformis*, *B. subtilis*, *B. pumilus* and *B. cereus* group organisms, as usually cited by papers reporting on microbial diversity in milk and derived products. Additionally, the potential of many isolates to produce lecithinases and/or proteases is shown by characteristic growth patterns on the differential media NAE and SM as well as their potential to grow at refrigeration temperatures (6 ± 1°C) after maximum 7 days. These findings underline again the hazardous implications spore-formers can have on milk quality and safety. Another important conclusion to be stressed out, is that about 7% of all isolates could not be attributed to already described species within *Bacillus* sensu lato, indicating our knowledge on spore-formers in milk (and to a larger extent in dairy products) is still far from complete, as already mentioned earlier.

It was found that total spore counts were higher in milk of conventional farms compared to organic farms, during summer months, and this difference could be mainly attributed to the higher occurrence of thermotolerant isolates (counts after incubation at 55°C). Previous studies also showed a seasonal variation for winter and summer total spore counts, with Christiansson *et al.* [8], Ellis *et al.* [12] and Lukasova *et al.* [35] reporting higher spore counts in summer due to lower udder hygiene and dietary changes. However, other studies indicate the opposite with higher spore counts in winter season due to housing [10, 56]. The thermotolerant isolates obtained in the comparative study of Coorevits *et al.* [9] were mainly identified as *B. licheniformis* and *Ureibacillus thermosphaericus*. Dietary differences could be a possible explanation for their higher occurrence in milk from conventional dairy farms. Indeed, cows on organic farms are fed locally cultivated crops, while cows on conventional dairy farms are fed maize, grass and wheat, supplemented with commercially available feed concentrate which is often composed of tropical (waste) ingredients. It is this feed concentrate in particular that is assumed to contribute to the higher thermotolerant spore counts in milk from conventional dairy farms. Indeed, the presence of thermotolerant isolates (e.g. *Bacillus licheniformis*) in feed concentrate has already been demonstrated by Vaerewijck *et al.* [63]. However, the presence of *U. thermosphaericus* in feed concentrate has not been shown. The only isolation source of *U. thermosphaericus* in the dairy industry reported thus far is milking equipment [51]. For that reason, one must be careful drawing conclusions and contamination sources other than feed concentrate should not be excluded. Another finding that possibly also could be attributed to these dietary differences is the higher diversity of isolates in milk from conventional dairy farms and the presence of more unidentified (onto the species level) isolates.

Another striking difference between milk from organic and conventional dairy farms was the more frequent occurrence of *B. cereus* group organisms in milk from organic dairy farms, and this was most obvious during summer months. As already mentioned, *B. cereus* is the most important pathogen for the dairy industry due to production of toxins causing food illnesses, production of deteriorative enzymes resulting in decreased milk quality and its ability to grow at storage temperatures (4-7°C) of pasteurised milk and other milk products. It is a common soil organism, and soiling of teats is thus most probably the major contamination source of raw milk with this pathogen. It was already shown that milk is more contaminated with *B. cereus* during summer period, when cows are more often grazing outside [8, 10]. Indeed, keeping cows indoor during summer months does decrease the presence of *B. cereus* [54]. The interview questionnaire performed during the study of Coorevits *et al.* [9] clearly indicated that cows from conventional farms were more often stabled indoor while cows from organic farms were grazing outside. So, differences in housing strategy may match with the observation that *B. cereus* group representatives are more frequently isolated from milk from organic dairy farms.

CONCLUSION

Quality and safety of milk is affected by the presence of contaminating bacteria, intruding raw milk at the milking stage, and hampering further production processes. Especially spore-formers are of concern, as the spores generated by these bacteria are resistant to the applied heat treatments and subsequently germinate into vegetative cell forms, spoiling milk via the production of deteriorating enzymes, toxins, nitrate reduction or gas production. Introduction of these spores into raw milk occurs via various contamination sources (feed, grass, air, faeces,...), and it can be concluded that differences in these contamination sources do affect the composition of the microbial flora in milk. Comparing the microbial load of milk from organic versus milk from conventional dairy farms clearly revealed a varying microbial constitution, which could most probably be attributed to differences in operational management between the two types of dairy farming.

More thermotolerant isolates (namely *Ureibacillus thermopshaericus* and *Bacillus licheniformis*) were found in milk from conventional dairy farms, which probably could be attributed to differences in feed strategy between conventional and organic farming. It is hypothesised that the use of feed concentrate in conventional dairy farms in particular contributes to this observation, as this feed concentrate is often composed of pelleted (hence heat-treated) and/or imported tropical waste ingredients. Thermotolerant bacteria are especially an issue in the production of milk powder. They often do not constitute a health risk to the consumer but are used as indicators to monitor good hygienic practices in milk powder production plants [48].

The higher occurrence of *Bacillus cereus* group organisms in milk from organic dairy farms compared to milk from conventional farms is another striking difference. This is most probably due to the housing strategy. For the animal's welfare, cows from organic farms are more often grazing outside, enabling a higher introduction of *B. cereus* spores into raw milk via grass and soil. As already mentioned *B. cereus* is the most important pathogen in the dairy industry because of its proteolytic and (phospho)lipolytic enzyme production causing structural defects in milk, its toxin production causing severe food illnesses and its ability to grow at refrigeration temperatures. Therefore, the presence of *B. cereus* is determinative for the shelf life of pasteurised milk and derived products.

Although differences in milk quality between milk from organic and conventional dairy farms have often been under study, the microbial diversity differences had not been investigated before to our knowledge. Most research was focused on protein and fat content of milk, as well as on the animal's welfare (occurrence of diseases). However, the results of our recent study [9] clearly indicate the correlation between operational management and microbial diversity of raw milk when focusing on aerobic spore-formers. In our opinion, an extension of this study to other contaminants and/or other countries is highly mandatory, as this would provide us with the necessary information about contamination sources, to take hygienic or other precautions in order to avoid, or at least reduce contamination risks at the initial phase, i.e. the milking stage.

REFERENCES

[1] Andersson, A; Ronner, U; Granum, PE. What problems does the food industry have with the spore-forming pathogens *Bacillus cereus* and *Clostridium perfringens*? *International Journal of Food Microbiology*, 1995, 28, 145-155.

[2] Barkley, MB; Delaney, P. An occurrence of bitty cream in Australian pasteurized milk. *Netherlands Milk and Dairy Journal*, 1980, 34, 191-198.

[3] Beattie, SH; Williams, AG. Detection of toxigenic strains of *Bacillus cereus* and other *Bacillus* spp. with an improved cytotoxicity assay. *Letters in Applied Microbiology*, 1999, 28, 221-225.

[4] Bennedsgaard, TW; Thamsborg, SM; Vaarst, M; Enevoldsen, C. Eleven years of organic dairy production in Denmark: herd health and production related to time of conversion and compared to conventional production. *Livestock Production Science*, 2003, 80, 121-131.

[5] Billing, E; Cuthbert, WA. "Bitty cream": The occurrence and significance of *Bacillus cereus* in raw milk supplies. *Journal of Applied Bacteriology*, 1958, 21, 65-78.

[6] Butler, G; Nielsen, JH; Slots, T; Seal, C; Eyre, MD; Sanderson, R; Leifert, C. Fatty acid and fat-soluble antioxidant concentrations in milk from high- and low-input conventional and organic systems: seasonal variation. *Journal of the Science of Food and Agriculture*, 2008, 88, 1431-1441.

[7] Chen, L; Coolbear, T; Daniel, RM. Characteristics of proteinases and lipases produced by seven *Bacillus* spp. isolated from milk powder production lines. *International Dairy Journal*, 2003, 14, 495-504.

[8] Christiansson, A; Bertilsson, J; Svensson, B. *Bacillus cereus* spores in raw milk: factors affecting the contamination of milk during the grazing period. *Journal of Dairy Science*, 1999, 82, 305-314.

[9] Coorevits, A; De Jonghe, V; Vandroemme, J; Reekmans, R; Heyrman, J; Messens, W; De Vos, P; Heyndrickx, M. Comparative analysis of the diversity of aerobic spore-forming bacteria in raw milk from conventional and organic dairy farms. *Systematic and Applied Microbiology*, 2008, 31, 126-140.

[10] Crielly, EM; Logan, NA; Anderton, A. Studies on the *Bacillus* flora of milk and milk products. *Journal of Applied Bacteriology*, 1994, 77, 256-263.

[11] Davies, FL; Wilkinson, G. *The microbiological safety of food. Bacillus in milk and dairy products*. Hobbs, BC; Christian, JHB (Ed.). Academia Press, London, United Kingdom. 1973.

[12] Ellis, KA; Innocent, GT; Mihm, M; Cripps, P; McLean, WG; Howard, CV; Grove-White, D. Dairy cow cleanliness and milk quality on organic and conventional farms in the UK. *Journal of Dairy Research*, 2007, 74, 302-310.

[13] Fagerlund, A; Ween, O; Lund, T; Hardy, SP; Granum, PE. Genetic and functional analysis of the *cytK* family of genes in *Bacillus cereus*. *Microbiology*, 2004, 150, 2689-2697.

[14] Foschino, R; Galli, A; Ottogalli, G. Research on the microflora of UHT milk. *Annals of Microbiology*, 1990, 40, 47-59.

[15] From, C; Hormazabal, V; Hardy, SP; Granum, PE. Cytotoxicity in *Bacillus mojavensis* is abolished following loss of surfactin synthesis: implications for assessment of toxicity and food poisoning potential. *International Journal of Food Microbiology*, 2007, 117, 43-49.

[16] From, C; Hormazabal, V; Granum, PE. Food poisoning associated with pumilacidin-producing *Bacillus pumilus* in rice. *International Journal of Food Microbiology*, 2007, 115, 319-324.

[17] From, C; Pukall, R; Schumann, P; Hormazabal, V; Granum, PE. Toxin-producing ability among *Bacillus* spp. outside the *Bacillus cereus* group. *Applied and Environmental Microbiology*, 2005, 71, 1178-1183.

[18] Granum, PE. Applications and Systematics of Bacillus and relatives. Bacillus cereus and food poisoning. Berkeley, RC; Heyndrickx, M; Logan, NA; De Vos, P (Ed.), Blackwell Science, Oxford, UK. 2002.

[19] Griffiths, MW; Phillips, JD. Incidence, source and some properties of psychrotrophic *Bacillus* spp. found in raw and pasteurized milk. *Journal of the Society of Dairy Technology*, 1990, 43, 62-66.

[20] Hammer, P; Lembke, F; Suhren, G; Heeschen, W. Characterization of a heat resistant mesophilic *Bacillus* species affecting the quality of UHT-milk - a preliminary report. *Kieler Milchwirtschaftliche Forschungberichte*, 1995, 47, 303-311.

[21] Hammer, P; Walte, HG. Zur pathogenität hitzeresistenter mesophiler sporenbilder aus UHT-milch. *Kieler Milchwirtschaftliche Forschungberichte*, 1996, 48, 151-161.

[22] Hardeng, F; Edge, VL. Mastitis, ketosis, and milk fever in 31 organic and 93 conventional Norwegian dairy herds. *Journal of Dairy Sciences*, 2001, 84, 2673-2679.

[23] Heyndrickx, M; Scheldeman, P. Applications and Systematics of Bacillus and relatives. Bacilli associated with spoilage in dairy and other food products. Berkely, R; Heyndrickx, M; Logan, NA; De Vos, P (Ed), Blackwell Science, Oxford, UK. 2002.

[24] Hoerning, B; Simantke, C; Aubel, E. *Investigations on dairy welfare and performance on German organic dairy farms*, 2005. Available from: http://orgprints.org/4371/.

[25] Klijn, N; Herman, L; Langeveld, L; Vaerewijck, M; Waegendorp, AA; Huemer, I; Weerkamp, AH. Genotypical and phenotypical characterization of *Bacillus* strains, surviving UHT sterilisation. *International Dairy Journal*, 1997, 7, 421-428.

[26] Klijn, N; Nieuwenhof, FF; Hoolwerf, JD; van der Waals, CB; Weerkamp, AH. Identification of *Clostridium tyrobutyricum* as the causative agent of late blowing in cheese by species-specific PCR amplification. *Applied and Environmental Microbiology*, 1995, 61, 2919-2924.

[27] Kondo, S; Sekine, J; Okubo, M; Asahida, Y. The effect of group size and space allowance on the agonistic and spacing behaviour of cattle. *Applied Animal Behaviour Science*, 1989, 24, 127-135.

[28] Labots, H; Hup, G. *Bacillus cereus* in raw and pasteurized milk. I. Enumeration of *B. cereus* spores in milk. *Netherlands Milk and Dairy Journal*, 1964, 19, 112-126.

[29] Labots, H; Hup, G; Galesloot, TE. *Bacillus cereus* in raw and pasteurized milk. III. The contamination of raw milk with *B. cereus* spores during its production. *Netherlands Milk and Dairy Journal*, 1965, 19, 191-221.

[30] Langford, FM; Rutherford, KMD; Jack, MC; Sherwood, L; Lawrence, AB; Haskell, MJ. A comparison of management practices, farmer-perceived disease incidence and winter housing on organic and non-organic dairy farms in the UK. *Journal of Dairy Research*, 2009, 76, 6-14.

[31] Lapidus, A; Goltsman, E; Auger, S; Galleron, N; Segurens, B; Dossat, C; Land, ML; Broussolle, V; Brillard, J; Guinebretiere, MH; Sanchis, V; Nguen-The, C; Lereclus, D; Richardson, P; Wincker, P; Weissenbach, J; Ehrlich, SD; Sorokin, A. Extending the *Bacillus cereus* group genomics to putative food-borne pathogens of different toxicity. *Chemico-Biological Interactions*, 2008, 171, 236-249.

[32] Lembke, F. Highly heat resistant spores in UHT-milk. *Bulletin no. 302. International Dairy Federation*, 1995.

[33] Lindsay, D; Mosupye, FM; Brozel, VS; Von Holy, A. Cytotoxicity of alkaline-tolerant dairy-associated *Bacillus* spp. *Letters in Applied Microbiology*, 2000, 30, 364-369.

[34] Llorens Abando, L; Rohner-Thielen, E. Different organic farming patterns within EU-25, an overview of the current situation. *Statistics in focus – Agriculture and Fisheries*, 2007, 69, (ISSN 1977-0316).

[35] Lukasova, J; Vyhnalkova, J; Pacova, Z. *Bacillus* species in raw milk and in the farm environment. *Milchwissenschaft*, 2001, 56, 609-611.

[36] McKellar, RC. (Ed.) *Enzymes of Psychrotrophs in Raw Food*. CRC Press, Inc., Boca Raton, Florida. 1989.

[37] Meer, RR; Bakker, J; Bodyfelt, FW; Griffiths, MW. Psychotrophic *Bacillus* spp. in fluid milk products: a review. *Journal of Food Protection*, 1991, 54, 969-979.

[38] Mikkola, R; Kolari, M; Andersson, MA; Helin, J; Salkinoja-Salonen, M.S. Toxic lactonic lipopeptide from food poisoning isolates of *Bacillus licheniformis*. *European Journal of Biochemistry*, 2000, 267, 4068-4074.

[39] Molkentin, J. Authentication of organic milk using $\delta^{13}C$ and the α-linolenic acid content of milk fat. *Journal of Agricultural and Food Chemistry*, 2009, 57, 785-790.
[40] Nauta, WJ; Baars, T; Bovenhuis, H. Converting to organic dairy farming: consequences for production, somatic cell scores and calving interval of first parity Holstein cows. *Livestock Science*, 2006, 99, 185-195.
[41] O'Donnovan, KO. The occurrence of *Bacillus cereus* in milk and on dairy equipment. *Journal of Applied Bacteriology*, 1959, 22, 131-137.
[42] Overcast, WW; Atmaram, K. The role of *Bacillus cereus* in sweet curdling of fluid milk. *Journal of Milk and Food Technology*, 1974, 37, 233-236.
[43] Palmer, J. Contamination of milk from the milking environment. *Kiel Milchwirtsch Forschungber*, 1981, 33, 307-316.
[44] Pettersson, B; Lembke, F; Hammer, P; Stackebrandt, E; Priest, FG. *Bacillus sporothermodurans*, a new species producing highly heat-resistant endospores. *International Journal of Systematic Bacteriology*, 1996, 46, 759-764.
[45] Phelps, RJ; McKillip, JL. Enterotoxin production in natural isolates of *Bacillaceae* outside the *Bacillus cereus* group. *Applied and Environmental Microbiology*, 2002, 68, 3147-3151.
[46] Phillips, JD; Griffiths, MW. Factors contributing to the seasonal variation of *Bacillus* spp. in pasteurized dairy products. *Journal of Applied Bacteriology*, 1986, 61, 275-285.
[47] Quiberoni, A; Guglielmotti, D; Reinheimer, J. New and classical spoilage bacteria causing widespread blowing in Argentinean soft and semihard cheeses. *International Journal of Dairy Technology*, 2008, 61, 358-363.
[48] Rueckert, A; Ronimus, RR; Morgan, HW. Development of a rapid detection and enumeration method for thermophilic bacilli in milk powders. *Journal of Microbiological Methods*, 2004, 60, 155-167.
[49] Salkinoja-Salonen, MS; Vuorio, R; Andersson, MA; Kampfer, P; Andersson, MC; Honkanen-Buzalski, T; Scoging, AC. Toxigenic strains of *Bacillus licheniformis* related to food poisoning. *Applied and Environmental Microbiology*, 1999, 65, 4637-4645.
[50] Scheldeman, P; Herman, L; Foster, S; Heyndrickx, M. *Bacillus sporothermodurans* and other highly heat-resistant spore formers in milk. *Journal of Applied Microbiology*, 2006, 101, 542-555.
[51] Scheldeman, P; Pil, A; Herman, L; De Vos, P; Heyndrickx, M. Incidence and diversity of potentially highly heat-resistant spores isolated at dairy farms. *Applied and Environmental Microbiology*, 2005, 71, 1480-1494.
[52] Schröder, MJA. Origin and levels of post pasteurization contamination of milk in the dairy industry and the effects on the keeping quality. *Journal of Dairy Research*, 1984, 51, 59-67.
[53] Shehata, TE; Duran, A; Collins, EB. Influence of temperature on the growth of psychrophilic strains of *Bacillus*. *Journal of Dairy Science*, 1971, 54, 1579-1582.
[54] Slaghuis, BA; te Giffel, MC; Beumer, RR; André, G. Effect of pasturing on the incidence of *Bacillus cereus* spores in raw milk. *International Dairy Journal*, 1997, 7, 201-205.
[55] Suominen, I; Andersson, MA; Andersson, MC; Hallaksela, AM; Kampfer, P; Rainey, FA; Salkinoja-Salonen, M. Toxic *Bacillus pumilus* from indoor air, recycled paper pulp, Norway spruce, food poisoning outbreaks and clinical samples. *Systematic and Applied Microbiology*, 2001, 24, 267-276.

[56] Sutherland, AD; Murdoch, R. Seasonal occurrence of psychrotrophic *Bacillus* species in raw milk, and studies on the interactions with mesophilic *Bacillus* sp. *International Journal of Food Microbiology*, 1994, 21, 279-292.

[57] Taylor, JM; Sutherland, AD; Aidoo, KE; Logan, NA. Heat-stable toxin production by strains of *Bacillus cereus, Bacillus firmus, Bacillus megaterium, Bacillus simplex* and *Bacillus licheniformis*. *FEMS Microbiology Letters*, 2005, 242, 313-317.

[58] te Giffel, MC; Beumer, RR; Slaghuis, BA; Rombouts, FM. Occurrence and characterization of (psychrotrophic) *Bacillus cereus* on farms in The Netherlands. *Netherlands Milk and Dairy Journal*, 1995, 49, 125-138.

[59] te Giffel, MC; Wagendorp, TA; Herrewegh, A; Driehuis, F. Bacterial spores in silage and raw milk. *Antonie van Leeuwenhoek*, 2002, 81, 625-630.

[60] Ternström, A; Lindberg, AM; Molin, G. Classification of the spoilage flora of raw and pasteurized bovine milk, with special reference to *Pseudomonas* and *Bacillus*. *Journal of Applied Bacteriology*, 1993, 75, 25-34.

[61] Tikofsky L. Milk *Quality on Organic Dairy Farms*. 2009. Available from: http://www.extension.org/article/18645.

[62] United States Department of Agriculture's Economic Research Service. Accessed March, 20, 2009. Available from: http://www.ers.usda.gov/.

[63] Vaerewijck, MJM; De Vos, P; Lebbe, L; Scheldeman, P; Hoste, B; Heyndrickx, M. Occurrence of *Bacillus sporothermodurans* and other aerobic spore-forming species in feed concentrate for dairy cattle. *Journal of Applied Microbiology*, 2001, 91, 1074-1084.

[64] Van Heddeghem, A; Vlaemynck, G. Sources of contamination of milk with *B. cereus* on the farm and in the factory. *Bulletin of the International Dairy Federation*, 1992, 275, 19-22.

[65] Vicini, J; Etherton, T; Kris-Etherton, P; Ballam, J; Denham, S; Staub, R; Goldstein, D; Cady, R; McGrath, M; Lucy, M. Survey of retail milk composition as affected by label claims regarding farm-management practices. *Journal of the American Dietetic Association*, 2008, 180, 1198-1203.

[66] Westhoff, DC; Dougherty, SL. Characterization of *Bacillus* species isolated from spoiled ultrahigh temperature processe milk. *Journal of Dairy Science*, 1981, 64, 572-580.

[67] Zwald, AG; Ruegg, PL; Kaneene, JB; Warnick, LD; Wells, SJ; Fossler, C; Halbert, LW. Management practices and reported antimicrobial usage on conventional and organic dairy farms. *Journal of Dairy Sciences*, 2004, 87, 191-201.

In: Organic Farming and Peanut Crops
Editors: D. C. Grossman, T. L. Barrios, pp. 137-151
ISBN: 978-1-60876-187-6

Chapter 5

THE EFFECT OF DISINFECTION ON SOIL BACTERIAL FUNCTIONALITY IN ORGANICALLY AND CONVENTIONALLY CULTIVATED PLOTS

P. D. Kapagianni and E. M. Papatheodorou*

Department of Ecology, School of Biology,
Aristotle Univ. Thessaloniki, Greece 54124

ABSTRACT

Bacterial functionality, as this was described by the utilization of substrates in the GN Biolog plates, was studied in different combinations of disinfection (chemical, biological, without disinfection) and cultivation (organic vs conventional) since both are common practices in agroecosystems. Within an area of 3000 m^2, which was conventionally cultivated with *Phaseolus vulgaris* (L.) for several years, a field plot experiment was set consisting of 18 plots (10x10 m each). Six plots were disinfected chemically with metham sodium, six were disinfected with organic additives (neem_cake and a mixture of basil oil, marigold and clove oil) while in the rest six plots no disinfectants were applied. Three plots from each disinfection type were cultivated organically receiving cow manure and the other three were cultivated conventionally. All plots were sowed by *Phaseolus vulgaris*. Bacterial functionality was examined in six treatments: chemical disinfection-organic cultivation (CD-O), biological disinfection-organic cultivation (BD-O), without disinfection–organic cultivation (WD-O), chemical disinfection-conventional cultivation (CD-C), biological disinfection-conventional cultivation (BD-C) and without disinfection-conventional cultivation (WD-C). Soil samples were collected at two sampling occasions: 1) at the flowering stage and 2) one month after the harvest of plants' aboveground biomass.

Shannon diversity index showed no response either to treatment or to sampling date reflecting the elimination of quantitative differences in bacterial functional diversity. On the contrary when the parameters K (maximum absorbance) and r (rate of changes from zero to K), which produced by the logistic curves fitted to the optical density data, were analysed, samplings and treatments were clearly separated. Since the distinction between treatments referred only to the degree of the utilization of some of the substrates, the

* Corresponding author: E-mail: papatheo@bio.auth.gr

differences in bacterial functionality were qualitative and not quantitative. The substrates responsible for the discrimination between treatments corresponded only to the 8-12% of the total_number of substrates used. It seemed that either cultivation or disinfection were disturbances of low severity for the bacterial functionality. The type of disinfection seemed to have a greater influence in conventional cultivated plots, by creating more distinct groups of samples, than in organically cultivated ones. We could hypothesized that the presence of this legumes species for several years in the study area, resulted in the dominance of a specific soil bacterial community, so the effects of the applied disturbances were of less importance for its function compare to those induced by the presence of this plant species.

INTRODUCTION

Mono-cultivated crops are easily affected by insects, nematodes, fungi, bacteria and weeds that compete with crops for nutrients. An effective practice to suppress soil borne pathogens (Hartz et al. 1989; Tu 1993; Welsh et al. 1997), nematodes (Harris 1991; Welsh et al. 1997), insects (Ware 1983; Weish et al. 1997) and weeds (McSorley et al. 1986; Jacobsohn et al. 1988; Welsh et al. 1997) is soil fumigation.

Metam sodium (sodium-N-methyldithiocarbamate) is a common fumigant widely used after the withdrawal of methyl bromide. It acts as a fungicide, herbicide, insecticide and nematicide simultaneously. Apart from chemical fumigants, which in some cases proved ineffective due to their quick degradation (Stromberger et al. 2005), there is an increasing tendency for using biocides friendlier to the environment. Such alternative biocides are the mixtures of essential oils and the products of neem tree as well. The latter has an antimicrobial activity (Converty and Allan 2001) whereas it suppresses nitrification (Parmar and Singh 1993; Prasad and Power 1995; Mohanty et al. 2008). Moreover, neem oil cake acts as an organic fertilizer that provides nutrients and improves the soil physico-chemical properties (Gopal et al. 2007).

Fumigants are among the pesticides with notable effects on soil microorganisms. The latter control nutrient availability and pesticide degradation with consequently effects on soil functioning and the productivity of agricultural systems (Klose and Ajwa, 2004; Klose et al, 2005). Taking into consideration that 80-90% of the processes in soil are reactions mediated by microbes (Coleman and Crossley, 1996; Nannipieri and Badalucco, 2003), the study of microbial diversity is of highly importance for agro-ecosystems' productivity. Most studies referred to functional microbial diversity because microorganisms that are present in soil are not necessarily functionally active (vanVeen and Heijnen, 1994; Meikle et al 1995; Trevors, 1998). The functional diversity includes the range and the relative expression of activities involved in such functions as decomposition, nutrient transformations, plant growth promotion or suppression and various other soil physical processes influenced by microorganisms (Giller et al 1997).

In order to estimate functional diversity the methods of SIR (Degens and Harris, 1997) and BIOLOG have been used. The first eliminates uncertainties due to extraction and culturing of microorganisms whereas the second one allows the comparison of microbial catabolic profiles across a large array of samples, rapidly, economically and effectively (Papatheodorou et al. 2008). In this work, we used the BIOLOG method primarily issued by Garland and Mills (1991). However, there are several drawbacks such as the no measurement of contribution of fungi because of their slow growth (Haack et al. 1995) and/or the possible changes in microbial community structure during the incubation period (Smalla et al. 1998).

Soil bacterial functional diversity had been studied in relation to management practices. The effects of no tillage, reduced and conservation tillage on functional diversity were investigated by Emmerling et al (2003), the influence of tillage and crop rotation by Lupwayi et al (1998), whereas Schutter et al (2001) examined the effects of cover cropping on microbial structure and function. Concerning the effects of chemical inputs on soils, the extent of disturbance that their application induced on the functionality of microbes was studied. Atlas et al (1991) examined the response of microbial populations after the addition of the herbicide 2,4,5-trichlorophenoxyacetic and Ibekwe et al (2001) applied fumigants to monitor the biocidal effects on soil microbial community.

This study focused on the response of microbial functional diversity when different types of disinfectants (chemical, biological) and fertilizers (organic, inorganic) were applied. According to Fujino et al. (2008) the incorporation of organic additives in the case of chemical disinfection could ameliorate the harmful effects of this latter regarding some of the soil enzymatic activities. The question posed herewith was how the two practices commonly used in agroecosystems -fertilization and disinfectation- affected the soil bacterial functionality and diversity and which was the significance of these effects compare to the temporal ones.

MATERIALS AND METHODS

Study Area

The study was conducted in Prespa region, located in the Northwestern part of Greece. It is a high altitude basin comprising two connected lakes (Micro-Prespa and Macro-Prespa), which are located at approximately 850 m asl and surrounded by forested mountains. The area covering the two lakes and the neighbouring slopes of Mounts Triklari and Varnunda is the largest National Park in Greece. The climate is mild continental-central European with Mediterranean features as well. The average annual precipitation ranges between 600 and 900 mm, and the wet season lasts from October to May. Snowfall is common from October until April (data from WWF-Greece, 2002).

Table 1. Physical soil properties of the experimental site.

Physical variables	Mean ± SE
Silt %	25.16 ± 1.88
Clay %	18.83 ± 1.07
Sand %	56 ± 2.43
pH	5.9 ± 0.064
Bulk densiity (g. cm^{-3})	1.63 ± 0.033

Experimental Design

Within an area of 3000 m^2, which was conventionally cultivated with shell beans (*Phaseolus vulgaris* L.) for several years, a field plot experiment was set consisting of 18 plots (10 x 10 m each). The soil physicochemical properties of the experimental area were presented in table 1.

Six plots were disinfected chemically with metham sodium (10 l NEMASOL 51 SL/100 l H_2O/100 m^2), six were disinfected with organic additives while in the rest six plots no disinfectants were applied. Organic additives consisted of neem cake (a product of the neem tree *Azadirachta indica* A. Juss) and a mixture of plant essential oils. The oil mixture consisted of basil oil, marigold and clove oil, and is recommended in organic farming to be used together with neem (www.agrosector.gr). Neem was dispersed by hand (6 kg/100 m^2) whereas the oil mixture was sprayed over the plots (0.33 l/100 l H_2O/100 m^2). Disinfectants were applied two months before the sowing of bean seeds. One and two months after disinfection, organic and inorganic fertilizers were applied to soils, respectively. Three plots from each disinfection type were cultivated organically receiving cow manure with an application rate of 2.5 tn / 1000 m^2 and the other three were cultivated conventionally receiving 8-16-24 (N-P-K) fertilizer. The latter was applied in a rate of 50 kg/ 1000 m^2. A day after the application of fertilizers the sowing of beans was carried out. All plots were randomly interspersed, while untreated empty spaces of 2 m width were left between them.

In total six combined treatments were created (3 types of disinfection x 2 types of cultivation) and these were: chemical disinfection - organic cultivation (CD-O), biological disinfection – organic cultivation (BD-O), without disinfection –organic cultivation (WD-O), chemical disinfection - conventional cultivation (CD-C), biological disinfection – conventional cultivation (BD-C), without disinfection –conventional cultivation (WD-C)

Soil Sampling

Sampling followed a randomized complete design involving three genuine replicate plots in each treatment (6 treatments x 3 replicate plots). To face intra plot variability, from each plot three soil samples were collected from the upper 12 cm of the soil. The diameter of each core was 7.5 cm. Samplings were conducted on two occasions related to different stages of crop development. The first one was in the stage of flowering at the beginning of August, and the second, one month after the harvest of the above-ground biomass, at the end of October. The 1st and 2nd sampling took place 4 and 7 months after disinfection, and about 3 and 6 months after the application of the fertilizers.

Bacterial Functional Diversity

For the determination of the bacterial functional diversity the method of Vahjen et al. (1995) was used. Samples were analysed during the first 10 days after their collection. During this time, they were kept in the refrigerator. Five grams from each soil sample were suspended for 30 min at 4^0C with 10 ml of sterile 0.85% NaCl solution at 220 rpm on a shaker. The soil slurry was sedimented for one h at 4^0C. Four millilitres of the suspension were removed and diluted with 12 ml of a sterile 0.85% NaCl solution. Each well of the Biolog-GN plates was inoculated with 100 ml of the above dilution. Bacterial substrate utilization was photometrically determined at 590 nm at 24 h intervals after incubation at 28^0C in the dark for 144 h. The absorption value of the control well was subtracted from each substrate absorption value while substrates with negative values were considered non-oxidized.

Data Analysis

To analyse the data, we used the mean value produced by the three replicate samples collected from each plot in order to have only genuine replicates. Thus, each treatment represented by three genuine replicates.

Individual values of optical density (OD) were grouped into 6 categories (polymers, carboxylic acids, carbohydrates, amino acids, amines/amides and miscellaneous). These data, along with values concerning diversity indices such as Shannon diversity and evenness were analyzed by means of a two-way ANOVA, with sampling date and treatment as independent variables.

Shannon diversity and evenness are described by the following equations:

$$\text{Diversity} = \Sigma\, p_i \ln(p_i) \quad \text{and}$$

$$\text{Evenness} = \ln\,(\text{diversity})\,/S$$

p_i = proportional color development of the ith well over total color development of all wells of a plate and S = number of wells with color development.

Data were primarily examined in order to meet the ANOVA assumptions (normality, homogeneity of variance etc.) and they were transformed whenever the assumptions were not met. In the case of significant effects, a Bonferonni test was further performed.

Apart from diversity parameters, individual BIOLOG recordings (OD of each well), were further analyzed. Specifically, on the successive readings of colour development of each well that were recorded every 24 h for an incubation period of 144 h, a sigmoid curve was fitted according to the equation proposed by Lindstrom et al. (1998):

$$OD_{595\ nm} = \frac{K}{(1 + e^{-r(t-s)})}$$

where K = maximum absorbance, r = the constant OD development rate, s = the time to the inflection point of the curve (K/2) and t = time. From the 95 substrates included in each plate, 85 were consumed in all samples and these were analyzed. Also, only K and r values were considered in this study since Lindstrom et al. (1998) mentioned that s parameter is correlated with inoculum size. They were analysed by using classification trees. Classification trees were primarily used to model ecological data by De'ath and Fabricius (2000). They use an algorithm in order to achieve repeatedly partitions of the data set into a nested serious of mutually exclusive groups, each of them as homogeneous as possible with respect to the predictor variable. All possible splits for each predictor variable at each node are examined to find the split producing the largest improvement in goodness of fit. K and r values were the continuous predictors in the current study, sampling was the categorical predictor whereas the six treatments inserted in the analysis as the dependent variables.

RESULTS

According to Table 2 the average optical density per plate and the mean OD of all substrate groups, except miscellaneous, were affected significantly by the sampling date. Their values were always higher in August (Table 3). Treatment had no effect in any of the studied parameters apart from evenness. Higher values of evenness were recorded in the plots where cow manure was incorporated as fertilizer (Figure 1). Plots under conventional cultivation exhibited lower values of evenness with the exception of WD-C plots. Diversity was unaffected either by sampling or by treatment whereas the interactive effect sampling x treatment proved insignificant for any of the parameters.

Classification trees applied on K and r values were presented in Figures 2 and 3. The classification tree based on K values was consisted of 12 terminal and 11 non-terminals nodes (Figure 2). Sampling that was inserted in the analysis as categorical predictor dichotomized the tree from its top by forming two branches with the same number of recordings (18 in each branch). The left one contained the data of August and the right one the data of October.

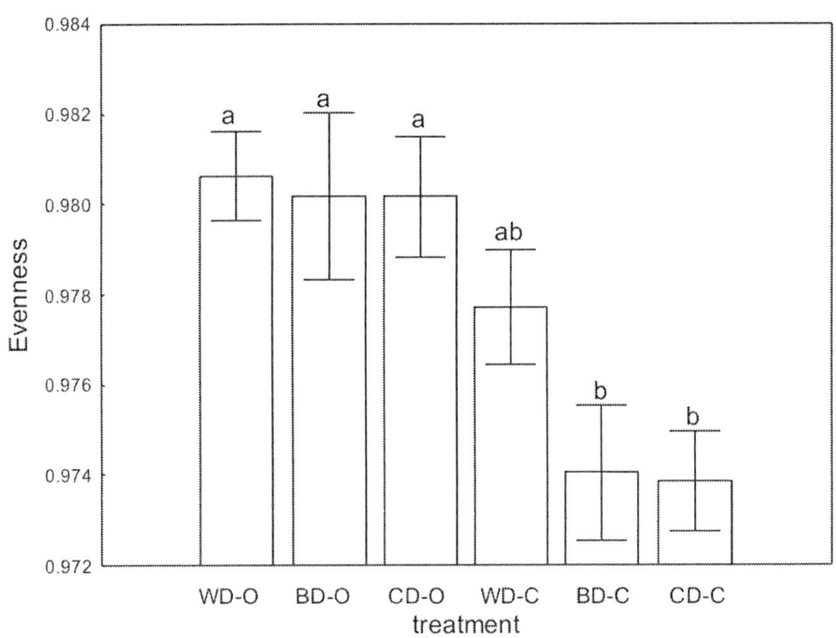

Figure 1. Evenness values in the six treatments (BD-C: biological disinfection-conventional cultivation, CD-C: chemical disinfection-conventional cultivation, WD-C: without disinfection-conventional cultivation, BD-C: biological-disinfection-conventional cultivation, CD-O: chemical disinfection-organic cultivation and WD-O: without disinfection-organic cultivation). Different letters correspond to statistically significant differences.

Table 2. The effect of sampling date (SD), treatment (TR) and their interaction on bacterial functional diversity and on mean catabolic activity (ns: non significant).

	SD (p-values)	TR (p-values)	SD x TR (p-values)
Mean OD per plate	0.00051	ns	ns
Diversity	ns	ns	ns
Evenness	ns	0.00055	ns
Carbohydrates	0.00000	ns	ns
Carboxylic acids	0.01493	ns	ns
Amino acids	0.03149	ns	ns
Amines/Amides	0.00221	ns	ns
Polymers	0.00045	ns	ns
Miscellaneous	ns	ns	ns

Table 3. Mean catabolic activity (± standard error) per plate and per substrate group in relation to sampling (a,b significant differences).

	Sampling time	
	August Mean ± SE	**October** Mean ± SE
Mean catabolic activity per plate	144.09±2.75[a]	130.33±2.43[b]
Carbohydrates	59.61±0.97[a]	52.99±0.89[b]
Carboxylic acids	27.67±0.69[a]	25.26±0.62[b]
Amino acids	31.33±0.68[a]	29.34±0.53[b]
Amines/amides	7.66±0.23[a]	6.71±0.17[b]
Polymers	9.43±0.22[a]	8.37±0.18[b]
Miscellaneous	8.37±0.3[a]	7.65±0.27[b]

In the sampling of August the K values of Glycerol, L-Proline and Thymidine contributed to treatments' classification. The WD-O treatment was clearly discriminated from all the others at the top of the left branch and it was characterized by the low K values of glygerol. On the contrary, the other organically cultivated treatments (BD-O and CD-O) were discriminated at the bottom of the branch and they were nearby to each other as well as nearby to plots treated conventionally. Specifically, the discrimination among the treatments BD-C and CD-C were due only to the utilization of glycerol. In the sampling of October, at the left branch of the tree, the substrates contributed significantly to the discrimination were L-Ornithine, L-Asparagine, α-D-Glucose, Bromosuccinic Acid and 2-Aminoethanol. It is remarkable that there was a successional discrimination of treatments cultivated conventionally from the top to the middle of the right branch (CD-C, BD-C and WD-C) whereas for those cultivated organically the discrimination was not as clear and it occurred at the bottom of the branch. From the organically cultivated treatments, CD-O was the most cohesive. On the contrary, in the WD-O treatment the misclassification error (number of cases that were not classified correctly) was 33%.

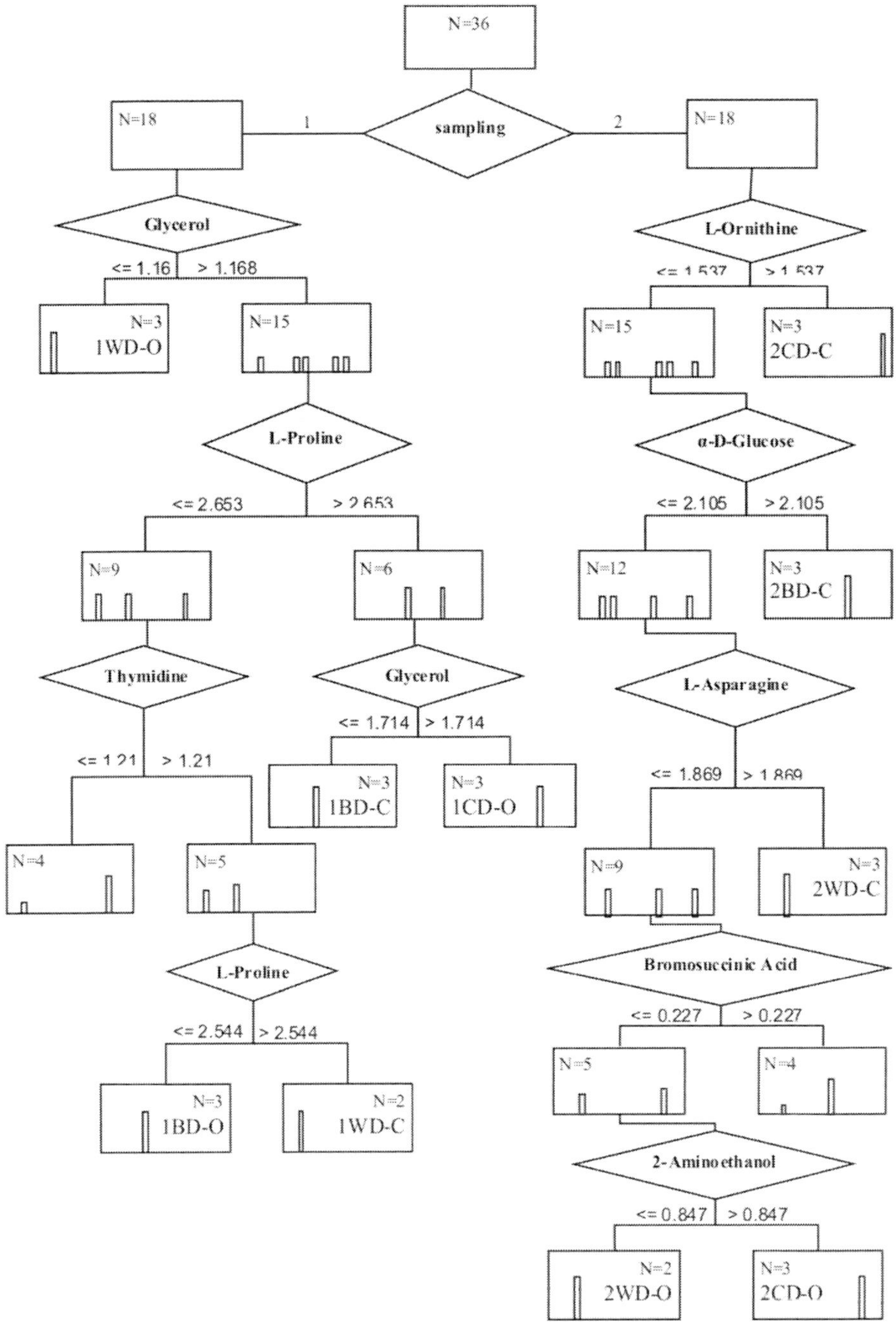

Figure 2. Classification tree based on K values. Inside nodes histograms represent treatments' frequencies. Each treatment is presented by an acronym where the number refers to sampling (1: August, 2: October), the first two letters to disinfection type (CD: chemical disinfection, BD: biological disinfection and WD: without disinfection) and the letter after the dash to the type of cultivation (C: conventional, O: organic).

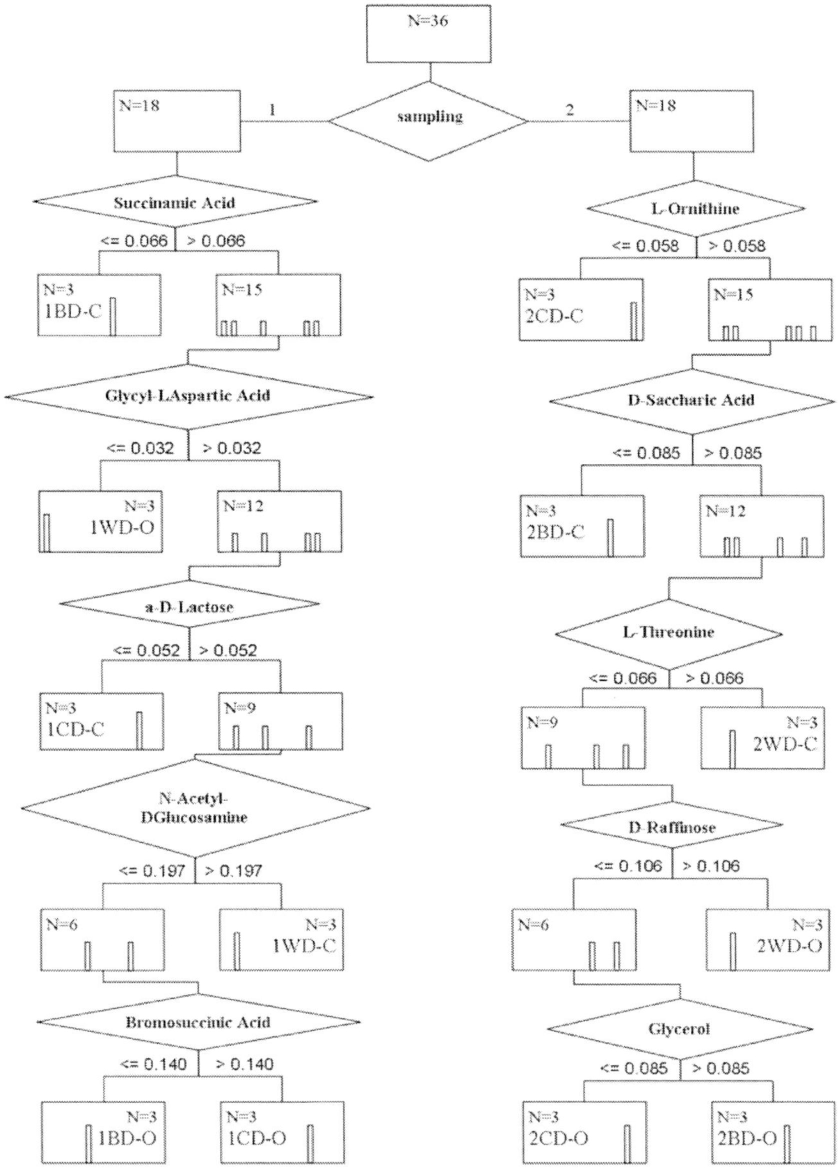

Figure 3. Classification tree based on r values. For details see Figure 2.

The classification tree that was constructed according to the r values had 12 terminal and 10 non-terminal nodes (Figure 3). Similarly to the previous tree, data was dichotomized by sampling, forming two branches of 18 recordings each one. The left branch referred to August and the right one to October sampling. In both samplings, the samples of each treatment constituted a cohesive group clearly separated from all others. The discrimination between treatments in August was owed to Succinamic Acid, Glycyl-LAspartic Acid, α-D-Lactose, N-Acetyl-D Glucosamine and Bromosuccinic Acid whereas in October the discrimination was owed to L-Ornithine, D-Saccharic Acid, L-Threonine, D-Raffinose and Glycerol. The pattern of discrimination was similar between samplings. The treatments BD-C, CD-C and WD-C were clearly separated from the top to the middle of the tree while the treatments BD-O and CD-O occupied terminal nodes at the bottom of the tree. The two latter treatments were separated due to the utilization of only one substrate.

Table 4. The substrates with the highest importance participating in treatments' separation according to K values

Substrates	Importance value
α-Keto Valeric Acid	1
Formic Acid	0.99
D-Mannitol	0.95
Maltose	0.91
L-Proline	0.87
D-Mannose	0.86
Succinic Acid Mono-Methyl-Ester	0.85
Glycerol	0.84
Lactulose	0.79

Table 5. The substrates with the highest importance participating in treatments' separation according to r values

Substrates	Importance value
D-Saccharic Acid	1
Bromosuccinic Acid	0.95
D-Arabitol	0.93
α-Cyclodextrin	0.93
α-D-Glucose	0.93
Glycyl-LAspartic Acid	0.93
β-Hydroxybutyric Acid	0.86
α-D-Lactose	0.86
Xylitol	0.85
Succinamic Acid	0.82
L-Threonine	0.81
Citric Acid	0.81
L-Serine	0.81
Glycerol	0.79

The relevant importance of substrates contributed in treatments' classification according to K and r values were presented in Tables 4 and 5, respectively. The importance of each predictor variable is estimated by summing - over all nodes in the tree - the drop in node impurity and expressing these sums relative to the largest sum found over all predictors. With this approach it is entirely possible to see importance values for predictors that were never chosen whereas it gives the opportunity for variables that may contain important predictive power with respect to the outcome of interest to be identified. The substrates with importance value ≥ 0.8 participated in classification tree based on K parameter were the following eight: Maltose, D-Mannitol, D-Mannose, Succinic Acid Mono-Methyl-Ester, Formic Acid, α-Keto

Valeric Acid, L-Proline and Glycerol. Similarly in treatments' classification based on r values the following substrates exhibited the highest importance: α-Cyclodextrin, D-Arabitol, α-D-Glucose, α-D-Lactose, Xylitol, Citric Acid, β-Hydroxybutyric Acid, D-Saccharic Acid, Bromosuccinic Acid, Succinamic Acid, Glycyl-LAspartic Acid, L-Serine, L-Threonine, Glycerol.

DISCUSSION

Functional diversity as this was estimated by the Shannon diversity index was not affected by sampling time. It was estimated at two sampling occasions coincided to different stages of crop development. Although diversity was unaffected, the mean catabolic activity per plate and the catabolic activity of the most substrates' groups differed significantly between samplings, exhibiting higher values in August. It seems that the composition of the soil bacterial community in terms of functional groups remain stable over time whereas a decrease in catabolic activity of all functional groups was recorded from August to October. The latter could be related mainly to the presence and absence of plants in August and October respectively, although the changes in abiotic conditions between these two periods could be also important. In August the plants, by producing roots' exudates, created a unique environment hospitable for specific groups of bacteria (Garbena et al 2004). In October, although plants were harvested, their roots remained into soil and because of their leguminous origin constitute important C and vitamins sources for bacteria (Altieri, 1999).

The application of different types of disinfection and fertilization seemed to have no effect on the diversity and the mean catabolic activity. Frequently, a quick recovery of the functional diversity after disturbance was reported in the literature. Ibekwe et al (2001) found an alteration of heterotrophic soil microbial communities by the application of fumigants during the first week of the experiment. Even in the case of burning, Ascoli et al (2005) reported a rapid recovery of functional diversity in soil whereas Calbrix et al (2007) concluded that C and lignin content in organic additives influenced the bacterial functional diversity during the first three months after their application. In the current study, samples were collected 4 and 7 months after disinfection and 3 and 6 months after fertilization and this could explain the elimination of significant effects. Different was the pattern of evenness index which affected significantly by treatments and specifically by fertilization. Higher evenness values were recorded in plots receiving cow manure (CD-O and BD-O), indicating a more equal utilization of all substrates.

A different approach of microbial functionality could be obtained by the study of parameters K, r and s since these latter were produced by considering the colour development of each well along the incubation period while the estimation of Shannon diversity based on one time data. Preston-Mafham et al (2002) advocate that kinetic parameters may be the most useful format for BIOLOG data, as they are only very weakly correlated with each other. As concerns the importance of the parameters, they suggested k parameter as the most useful. Moreover, Lindstrom et al. (1998) reported that both K and r parameters are constant under stable temperature incubation conditions and they depend on the taxonomic richness and the relative abundance of microplate-culturable microorganisms and not on the inoculums density per se, as occurs with the parameter s.

In both trees and sampling periods, the treatments BD-C and CD-C were differentiated at the top of the trees, contrary to the treatments BD-O and CD-O that were differentiated at the bottom. This fact showed that the type of disinfection (biological - chemical) had a significant influence in conventionally cultivated plots resulting in bacterial communities with distinct characteristics. On the contrary, the bacterial communities in BD-O and CD-O treatments were not as distinguishable as the above mentioned ones. It is probably related to the fact that in BD-O and CD-O plots two types of disturbances were involved; the disinfection and the conversion of a long-term conventionally cultivated area to an organic one. On the contrary, in BC-C and CD-C plots the disturbance was only that of disinfection.

Treatments' separation was due to the utilization of substrates belonging to the groups of amines/amides, amino-acids and miscellaneous whereas those belonging to most numerous groups such as carbohydrates and carboxylic acids did not participate. Furthermore, although the construction of k and r trees based on the utilization of 85 substrates, only 8 and 14, respectively, were the most important for treatments' separation. It seemed that the qualitative differences between treatments were focalized on a short array of substrates. We could hypothesized that this plant species which has been cultivated conventionally for years in the experimental area, it has developed a specific soil bacterial community in terms of function which changed unremarkable under disinfection and fertilization treatments. We concluded that the type of disinfection and cultivation involved in this study did not constitute powerful pressures for microbial functionality.

Conclusion

In conclusion, disinfection and fertilization were not considered severe disturbances for the functionality of the soil bacterial community. The differences in well consumption between treatments were qualitative and they referred to the utilization of a small number of substrates. On the contrary, the differences between samplings were both quantitative, as this revealed by the differences in the mean catabolic activity, and qualitative. It seemed that the temporal effect was more pronounced for the functionality of the soil microbial community than the effect of treatments. In this line of evidence was the study of Papatheodorou et al. (2008) which demonstrated the dominance of temporal changes in the composition of consumed substrates compared to changes induced by the different management regimes.

Acknowledgments

This work was part of the project "Use of alternative methods for improving soil fertility" (PENED 2004), funded by the General Secretariat of Research and Technology (Greek Ministry of Development).

References

Altieri, M. A. (1999). The ecological role of biodiversity in agroecosystems. *Agriculture, ecosystems and environment*, *74*, 19-31.

Atlas, R. M., Horowitz, A., Krichevsky, M. & Bej, A. K. (1991). Response of Microbial Populations to Environmental Disturbance. *Microbial ecology*, *22*, 249-256.

Calbrix, R., Barray, S., Chabrerie, O., Fourrie, L. & Laval, K. (2007). Impact of organic amendments on the dynamics of soil microbial biomass and bacterial communities in cultivated land. *Applied soil ecology*, *35*, 511-522.

Coleman, D. C., Crossley, D. A. & Hendrix, P. F. (1996). *Fundamentals of soil ecology* (2nd edition). London, Academic press.

Coventry, E. & Allan, E. J. (2001). Microbiological and chemical analysis of neem (Azadirachta indica) extracts: new data on antimicrobial activity. *Phytoparasitica*, *29*, 441-450.

D'Ascoli, R, Rutigliano, F. A., De Pascale, R. A., Gentile , A. & De Santo, A. V. (2005). Functional diversity of the microbial community in Mediterranean maquis soils as affected by fires. *International journal of wildland fire*, *14*, 355-363.

De'ath, G. & Fabricius, K. E. (2000). Classification and regression trees: a powerful yet simple technique for ecological data analysis. *Ecology*, *81*, 3178-3192.

Degens, B. P. & Harris, J. A. (1998). Development of a physiological approach to measuring the catabolic diversity of soil microbial communities. *Soil biology and biochemistry*, *29*, 1309-1320.

Emmerling, C., Gattinger, A. & Embacher, A. (2003). *Reduced tillage in organic farming: impact on microbial activity and functional and structural diversity of soil microbial community*, http://orgprints.org/00001736. emmerling@uni-trier.de

Fujino, C., Wada, S., Konoike, T., Toyota, K., Suga, Y. & Ikeda, J. (2008). Effect of different organic amendments on the resistance and resilience of the organic matter decomposing ability of soil and the role of aggregated soil structure. *Soil science and plant nutrition*, *54*, 534-542.

Garbeva, P., van Veen, J. A. & van Elsas, J. D. (2004). Microbial diversity in soil: selection of microbial populations by plant and soil type and implications for disease suppressiveness, *Annual review of phytopathology*, *42*, 243-270.

Garland, J. L. & Mills, A. L. (1991). Classification and characterization of heterotrophic microbial communities on the basis of patterns of community-level sole-carbon-source utilization. *Applied and environmental microbiology*, *57*, 2351-2359.

Giller, K. E., Beare, M. H., Lavelle, P., Izac, A. M. N. & Swift, M. J. (1997). Agricultural intensification, soil biodiversity and agroeco-system function. *Applied soil ecology*, *6*, 3-16.

Gopal, M., Gupta, A., Arunachalam, V. & Magu, S. P. (2007). Impact of azadirachtin, an insecticidal allelochemical from neem on soil microflora, enzyme and respiratory activities. *Bioresource technology*, *98*, 3154-3158.

Haack, S. K., Garchow, H., Klug, M. J. & Forney, L. J. (1995). Analysis of factors affecting the accuracy, reproducibility, and interpretation of microbial community carbon source utilization patterns. *Applied and environmental microbiology*, *61*, 1458-1468.

Harris, D. C. (1991). A comparison of dazomet, chloropicrin and methyl bromide as soil disinfestants for strawberries. *Journal of horticultural science*, *66*, 51-58.

Hartz, T. K., Bogle, C. R., Bender, D. A. & Avila, F. A. (1989). Control of pink root disease in onion using solarization and fumigation. *Journal of american society for horticultural science*, *114*, 587-590.

Ibekwe, A. M., Papiernik, S. K., GAN, J., Yates, S. R., Yang, C. H. & Crowley D. E. (2001). Impact of fumigants on soil microbial communities. *Applied and environmental microbiology*, *67*, 3245-3257.

Jacobsohn, R., Kelman, Y., Shaked, R. & Klein, L. (1988). Broomrape (Orobanche spp.) control with ethylene dibromide and chloropicrin. *Weed science*, *28*, 151-157.

Klose, S. & Ajwa, H. A. (2004). Enzyme activities and microbial biomass in agricultural soils fumigated with methyl bromide alternatives. *Soil biology and biochemistry*, *36*, 1625-1635.

Klose, S., Acosta-Martinez, V. & Ajwa, H. A. (2006). Microbial community composition and enzyme activities in a sandy loam soil after fumigation with methyl bromide or alternative biocides. *Soil biology and biochemistry*, *38*, 1243-1254.

Lindstrom, J. E., Barry, R. P. & Braddock, J. F. (1998). Microbial community analysis: a kinetic approach to contrasting potential C source utilization patterns. *Soil biology and biochemistry*, *30*, 231-239.

Lupwayi, N. Z., Rice, W. A. & Clayton, G. W. (1998). Soil microbial diversity and community structure under wheat as influenced by tillage and crop rotation. *Soil biology and biochemistry*, *30*, 1733-1741.

McSorley, R., McMillan, R. T. & Parrado, J. L. (1986). Comparative control of soilborne pests on tomato and pepper by soil fumigation. *Proceedings of the Florida State Horticultural Society*, *99*, 350-353.

Meikle, A., Amin-Hanjani, S., Glover, L. A., Killham, K. & Prosser, J. I. (1995). Matric potential and the survival and activity of a Pseudomonas fluorescens inoculum in soil. *Soil biology and biochemistry*, *27*, 881-892.

Mohanty, S., Patra , A. K. & Chhonkar, P. K. (2008). Neem (Azadirachta indica) seed kernel powder retards urease and nitrification activities in different soils at contrasting moisture and temperature regimes. *Bioresource technology*, *99*, 894-899.

Monokrousos, N., Papatheodorou, E. M. & Stamou, G. P. (2007). The response of soil biochemical variables to organic and conventional cultivation of Asparagus sp. *Soil biology and biochemistry*, *40*, 198-206.

Nannipieri, P. & Badalucco, L. (2003). Biological processes. In: D. K. Benbi, & R. Nieder, (Eds), Handbook of processes and modelling in the soil–plant system: *Modelling concepts and applications*, (57-82). Binghamton, The Haworth press.

Papatheodorou, E. M., Efthimiadou, E. & Stamou, G. P. (2008). Functional diversity of soil bacteria as affected by management practices and phonological stage of Phaseolus vulgaris. *European journal of soil biology*, *44*, 429-436.

Parmar, B. S. & Singh, R. P. (1993). *Neem in Agriculture*. New Delhi, Indian agriculture research institute.

Prasad, R. & Power, J. F. (1995). Nitrification inhibitors for agriculture, health and the environment. *Advances in agronomy*, *54*, 233-281.

Preston-Mafham, J., Boddy, L. & Randerson, P. F. (2002). Analysis of microbial community functional diversity using sole-carbon-source utilization profiles – a critique. *FEMS microbiology ecology*, *42*, 1-14.

Schutter, M. & Dick, R. (2001) Shifts in substrate utilization potential and structure of soil microbial communities in response to carbon substrate. *Soil biology and biochemistry*, *33*, 1481-1491.

Smalla, K., Wachtendorf, U., Heuer, H., Liu, W. & Forney, L. (1998). Analysis of Biolog GN substrate utilization patterns by microbial communities. *Applied and environmental microbiology*, *64*, 1220-1225.

Stromberger, M. E., Klose, S., Ajwa, H., Trout, T. & Fennimore, S. (2005). Microbial populations and enzyme activities in soils fumigated with methyl bromide alternatives. *Soil science society of America journal*, *69*, 1987-1999.

Trevors, J. T. (1998). Bacterial biodiversity in soil with an emphasis on chemically contaminated soils. *Water, air and soil pollution*, *101*, 45-67.

Tu, C. M. (1993). Effect of nematicides, Telone II & Vorlex, on microflora and nitrification in tobacco soil. *Bulletin of environmental contamination and toxicology*, *50*, 43-48.

van Veen, J. A. & Heijnen, C. E. (1994). The fate and activity of micro-organisms introduced into soil. In: C. E., Pankhurst, B. M., Doube, V. V. S. R. Gupta, & P. R. Grace, (Eds), *Soil biota: management in sustainable farming systems*, (63-71). Adelaide, CSIRO Australia.

Ware, G. W. (1983). Pesticides: theory and application. San Francisco, W. H. Freeman and company.

Welsh, C. E., Guertal, E. A. & Wood, C. W. (1998). Effects of soil fumigation and N source on soil inorganic N and tomato growth. *Nutrient cycling in agroecosystems*, *52*, 37-44.

In: Organic Farming and Peanut Crops
Editors: D. C. Grossman, T. L. Barrios, pp. 153-162 ISBN: 978-1-60876-187-6

Chapter 6

ABIOTIC STRESS AND RESPONSE OF PEANUT-RHIZOBIA

Marta Dardanelli, Victor Busto, Paola Pereira, Miguel Bueno and Mirta García

Departamento de Biología Molecular, Facultad de Ciencias Exactas, Físico-Químicas y Naturales, UNRC-Río Cuarto X5804BYA-Argentina

ABSTRACT

Peanut (*Arachis hypogaea* L.) is a regional crop of Argentina. It is grown in the centre-south of Córdoba province, where almost 96% of the national production and all of its industrial processing are concentrated (Fernandez and Giayetto, 2006). Peanut is an important subsistence and cash crop in the semi-arid tropics where it often suffers from drought or saline stress. Little is known about the molecular events involved in its adaptive responses to stress and about the biochemical and physiological basis of tolerance by rhizobia nodulating peanut roots and their symbiosis. The aim of the present chapter is to contribute to the knowledge of physiological and biochemical response (trehalose content; lipids and periplasmic glucans) of peanut microsimbionts under different types of stress. This response will be compared between rhizobia, and it probably will contribute in the selection of a great tolerant strain for future essays of inoculation.

Results indicate that there are different responses of peanut-rhizobia to abiotic stress, indicating genetic variability in tolerance to these environmental constraints. Trehalose accumulation is strain dependent. However other parameters were similar in different strains. Thus the polymerization degree of periplasmic glucans was affected under the stress conditions applied; this was evidenced in its molecular weight profile. Also phospholipids increased under hypersaline and hyperthermic conditions.

Results suggest that periplasmic glucans and phospholipids would be implicated in the adaptive response of peanut strains. Further analysis of these molecules in peanuts rhizobia must be carried out to select bacteria tolerant to stress for inoculants formulations.

INTRODUCTION

Bacteria inhabit natural and artificial environments with diverse and fluctuating physical properties, including osmolality, salinity and temperature. Although most, if not all, microorganisms use this strategy, the types of molecules accumulated and the accumulation mechanism (s) differ among species (Wood et al., 2001).

The rhizosphere, the soil zone influenced by the plant roots, is dynamic, and rhizobacteria may sense and respond to plant signals to exchange nutrients with plant cells, suffer damage due to plant defense response, and colonize or even invade root tissues, creating pathologies or symbiosis. As the activities of roots and rhizobacteria modify the osmolality of the rhizosphere, osmoadaptative mechanisms will likely influence both the survival of rhizosphere bacteria and their interactions with plant roots (Miller and Wood, 1996).

Rhizobia are able to fix atmospheric nitrogen in symbiosis with legumes; they represent a way toward environmentally sustainable agriculture. Rhizobacteria are thus economically important not only because of increase in agricultural legume yields, but also for the enrichment of the soil with nitrogen for the other crops and of the protection of soil and water from pollution caused by chemical fertilizers (Bouhmahdi et al., 1999).

The most problematic environments for rhizobia are marginal lands with low rainfall, extremes of temperature, acidic soil of low nutrient status, and poor water-holding capacity, and different species vary in their tolerance to major environmental factors (Zahran, 1999).

Rhizobia can adapt to saline conditions by the intracellular accumulation of low-molecular-weight organic solutes called osmolytes (Csonka and Hanson, 1991). Low molecular weight saccharides (LMWS) and disaccharide trehalose plays a role in osmoregulation when rhizobia are growing under salt or osmotic stress (El-Sheikh and Wood, 1990; Zorreguieta et al., 1990) or accumulates to higher levels in cells of *Rhizobium leguminosarum* (Breedveld et al., 1991) and peanut rhizobia (Dardanelli et al., 2000) under increasing osmotic pressure of hypersalinity.

One of the more important environmental factors influencing the activity of microorganisms is the soil water potential (Halverson and Firestone, 2000). Studies of bacterial water stress physiology have generally examined the genetic and physiological mechanisms of adaptation to osmotic stress caused by permeating solutes used to lower the water potential of the growth medium (Csonka, 1989). In many non saline soils, lowering of the water potential is due to a reduction in the water content and not to an increase in the concentration of permeating solutes, for this high molecular weight polyethylene glycol (PEG) has been used in plant and microbial studies related to responses to reduced water content, feigning environmental conditions (Busse and Bottomley, 1989).

The dehydration of membranes at a constant temperature causes a phase transition from liquid crystalline phases to a gel phase, that is, dehydration results in an increase in the transition temperature. Sucrose and trehalose can depress the phase transition temperature during desiccation and may contribute to the ability of many microorganisms to survive desiccation by maintaining the fluidity of the membrane (Potts, 1994).

Among the critical biochemical and physiological alterations that must by performed during fluctuating environmental conditions are alterations in the membrane lipid composition (DiRusso and Nyströn, 1998).

The phospholipid (PL) composition of the membrane may change in response to salt in the head-group of the lipids (Russell, 1995). Cardiolipin (CL) controls the osmotic stress response and the subcellular location of transporter ProP in *Escherichia coli*. (Romantsov et al., 2008).

Bradyrhizobium SEMIA 6144, a rhizobio peanut, is able to synthesize fatty acids, to incorporate them into its PL, and then modify them in response to stress conditions such as temperature and salinity (Medeot et al., 2007).

The PL of biological lipid bilayers are normally hydrated and disaccharides such as trehalose preserve the structural integrity of membranes in the dried state (Welsh, 2000). However the effect of different environmental stresses on rhizobia nodulating peanut is not yet well understood. The aim of the present work is to contribute to the knowledge of physiological and biochemical response (trehalose content; lipids and periplasmic glucans) of peanut microsimbionts under different types of stress. This response will be compared between rhizobia, and it probably will contribute in the selection of a great tolerant strain for future essays of inoculation.

Growth Characteristics under Stress Conditions

The strain, *Bradyrhizobium* TAL169, was grown under control conditions (28ºC), hypersalinity (100 mM NaCl), and hyperthermia (37ºC). Another peanut symbiont, *Bradyrhizobium* TAL1000, was used for comparative studies of hyperthermia stress. For this study we employed optical density (600nm), colony format unit recount (CFU/ml), wet weight, protein and trehalose quantification, periplasmic glucans profile and lipids determination.

When the viability of *Bradyrhizobium* TAL 169 was tested, it was observed that it was negatively affected by both, hyperthermia (43%) and PEG6000 16%P/V (52%), while it was not affected by ionic stress (Table 1). The generation times were affected in relation to the type of imposed stress and they correspond to the number of viable cells.

Table 1. Effects of abiotic stress on growth and trehalose content of *Bradyrhizobium* TAL169

Condition Parameters	Control 28ºC	Hyperthermia 37ºC	NaCl 100 mM	NaCl 400 mM	PEG-6000 16% p/v
Viability (CFU/ ml medium)	$6.50 \pm 0.32 \times 10^{9(a)}$	$3.72 \pm 0.33 \times 10^{9(b)}$	$7.35 \pm 0.47 \times 10^{9(a)}$	$6.63 \pm 0.12 \times 10^{9(a)}$	$3.15 \pm 0.30 \times 10^{9(b)}$
Biomass (mg/ml medium)	$0.17 \pm 0.02^{(a)}$	$0.51 \pm 0.02^{(b)}$	$0.63 \pm 0.04^{(b)}$	$1.43 \pm 0.20^{(c)}$	$0.315 \pm 0.038^{(b)}$
Proteins (mg/mg biomass)	$0.39 \pm 0.01^{(a)}$	$0.280 \pm 0.005^{(b)}$	$0.60 \pm 0.10^{(c)}$	$0.62 \pm 0.11^{(c)}$	$0.294 \pm 0.087^{(b)}$
Trehalose (µmol/mg proteins)	$0.178 \pm 0.022^{(a)}$	$0.142 \pm 0.013^{(a)}$	$0.030 \pm 0.008^{(b)}$	$0.015 \pm 0.003^{(c)}$	$0.352 \pm 0.041^{(d)}$

Dates are the means of three independent experiments. Significant differences ($p<0.05$) between values within a column are indicated by different letters.

These results are in agreement with those obtained by Kulkarni and Nautiyal (2000). Changes in growth rates of rhizobia have also been reported to be related to adaptation to different environmental situations such as pH, salt, and temperature (Correa et al., 1999; Drouin et al., 2000; Shamseldin and Werner, 2005). When the viability of TAL1000 was tested under hyperthermia, it was observed that it was not affected. However when the study was determinate in hypersalinity (100 mM NaCl)-hyperthermia (37ºC), viable cells suffered an increase (Figure 1). Similar results have been published by Mohammad et al. (1991), who noted a differential response of *Esinfer meliloti* to salinity and osmotic stress, indicating genetic variability in tolerance to these environmental constraints.

A

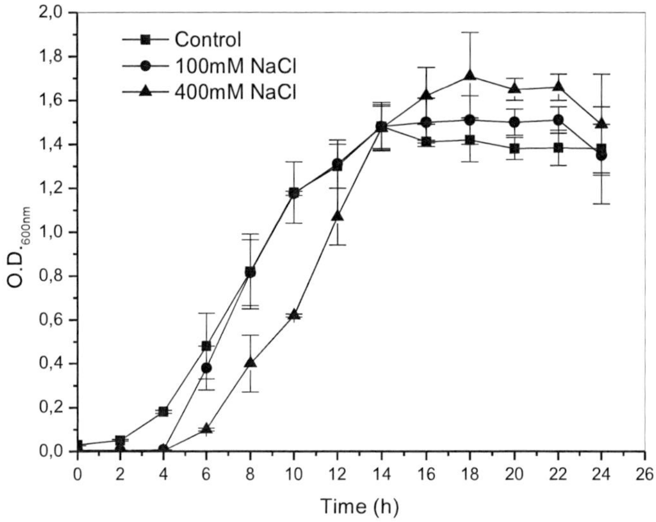

B

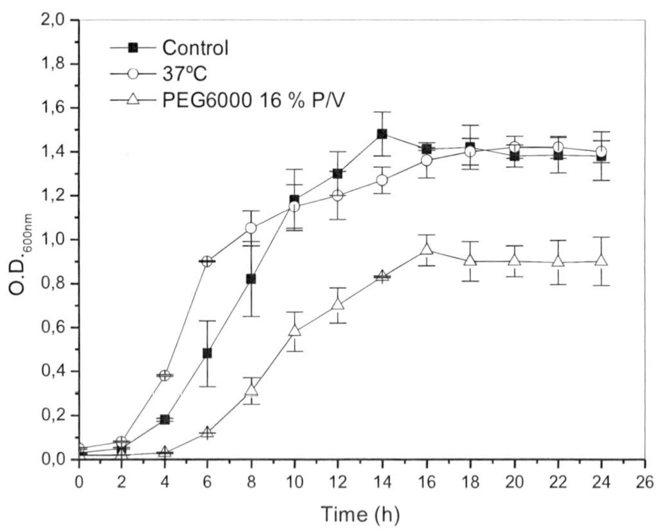

Figure 1. Effect of NaCl (A) and temperature y PEG6000 16%P/V (B) on *Bradyrhizobium* TAL169. Growth was measured by the increase in absorbance at 620 nm. Values represent means ± S.E.M. of three independent experiments.

Effects of Stress on the Content of Intracellular Solutes

Trehalose and sucrose have been shown to preserve both structure and function of isolated proteins during stress (Carpenter et al., 1987; Crowe et al., 1987). This ability to stabilize proteins during drying results from the disaccharides forming hydrogen bonds with the proteins when water is removed, thus preventing protein denaturation (Carpenter and Crowe, 1988). Adding trehalose before drying lowers the transition temperature (*Tm*) of the dry membranes by replacing the water between the lipid head-groups, preventing the phase transition and its accompanying leakage upon rehydration (Crowe et al., 1987).

When this disaccharide was quantified in *Bradyrhizobium* TAL169, it was observed that its content diminished 5.9 times and 11.8 times in the presence of 100 and 400 mM of NaCl respectively, with respect to the control value (Table 1).

In response to other types of stress, as nonionic stress, its concentration increased 2 times, whereas in hyperthermia significant changes in their value were not observed. In *Bradyrhizobium* TAL1000, the trehalose content diminished 1.6 times (from $1.33x10^{-2}$ to $0.8x10^{-2}$ μmol trehalose/mg proteins) with hyperthermia and increased 2 times (from $1.33x10^{-2}$ to $2.7x10^{-2}$ μmol trehalose/mg proteins) with hyperthermia-salinity. When this strain was dealt only with saline stress it was not observed change in the concentration of the osmoprotector (Dardanelli et al., 2009).

Trehalose has been shown to protect cell membranes and proteins from inactivation or denaturation caused by a variety of stress conditions (McIntyre et al., 2008). Previous work from our laboratory suggests that fast-growing peanut rhizobia accumulate trehalose in the presence of different carbon sources (mannitol, sucrose or lactose**).** It has been reported that slow-growing peanut rhizobia accumulate trehalose only when cultured with mannitol as the carbon source (Ghittoni and Bueno, 1995).

In osmotic and saline stress, peanut rhizobia ATCC10317, TAL1000, TAL1371 and SEMIA6144 showed different response of potassium and trehalose content. An explanation for this response can be the genetic variability of peanut simbionts or the use of other compatible solutes (Dardanelli et al., 2009).

Effects of Stress on the Phospholipids Content

The lipid bilayer forms the framework of the cytoplasmic membranes. The major lipid components of the bilayer are PL. They provide the membrane with its barrier function and play a role in a variety of processes in the bacterial cell.

Hyperthermia increased the content of PL of *Bradyrhizobium* TAL169 in 156% with respect to the control. Nevertheless, ionic stress did not affect this parameter unlike the nonionic stress that increased its content in a 238%

For the other strain studied, *Bradyrhizobium* TAL1000, variations were observed although smaller, thus, 21% was the increase of total PL when the applied stress was hyperthermia, whereas, an increase of 12% was caused by hipersalinidad-hyperthermia.

The concentration of each one of the individual phospholipids also was determined for both strains, with the intention of knowing if applied stress modifies or, the pattern of PL or its percentage of distribution. Table 2 shows the results corresponding to both parameters for *Bradyrhizobium* TAL169. It can be observed that the pattern of phospholipids was not modified, nevertheless were observed modifications in the percentage of distribution of the same. The most relevant changes were detected concerning the denominated PL fraction called phosphatidylglycerol + phosphatidylethanolamine (PG+PE) whose proportion diminished a 40%. Concomitantly with that diminution, hyperthermia caused an increase of the 185% in denominated dimethylethanolamine (DM) fraction.

Table 2. Effect of hyperthermia on phospholipid composition of *Bradyrhizobium* TAL169

	28ªC		37ºC	
PL	μg PL	% [a]	μg PL	% [a]
PC	3.91 ± 0.20	18.65	2.08 ± 0.35	20.88
DM	1.73 ± 0.16	8.27	2.35 ± 0.35	23.59
PG+PE	8.21 ± 0.92	39.17	2.36 ± 0.66	23.69
PA	3.54 ± 0.36	17.00	1.82 ± 0.71	18.27
CL	3.57 ± 0.21	16.90	1.35 ± 0.12	13.55

Bradyrhizobium TAL169 was grown in YEM medium and harvested at late exponential phase. Values represent means ± S.E.M. of tree independet experiments.
[a] Indicate each value with respect to the total, considered as 100%.

The pattern of PL of *Bradyrhizobium* TAL1000 was not modified either by applied stresses, but like for *Bradyrhizobium* TAL169, differences in the percentage distribution were observed. For this strain it was found variations concerning phosphatidylcholine (PC), zwitterionic phospholipid, whose proportion diminished a 48%, concerning phosphatidic acid (PA)+PG, anionic phospholipids, that increased a 40% and concerning CL, also an anionic phospholipid, that increased a 43%, all of them with respect to the control.

The previous thing it is come off that applied stresses influenced of particular way each one of the strain of studied rhizobia although they have in common that they affected the percentage distribution of PL for both strains. These findings agree with previous results of our laboratory using *Bradyrhizobium* sp strain SEMIA6144 where the most significant change associated with the stresses was a sustained increase in the total amount of PL with NaCl (38%) and a decrease with temperature stress (40%) and a 36% when both stresses were applied together (Medeot et al., 2007). These results taken altogether indicate that stress, saline, by temperature or non ionic can affect the polar head of the phospholipids in a differential way. These changes maintain the optimal physical state of the lipidic matrix necessary for membrane normal function of the bacterial cell (Suutari and Laakso, 1994).

The Effects of Abiotic Stress on Cell-Associated Low Molecular Weight Saccharides from Peanut Rhizobia Strains

Low molecular weight saccharides, most of them cyclic glucans, are present in the periplasmic space of rhizobia, where they play a role in osmotic adaptation (Dardanelli et al., 1997). When LMWS from *Bradyrhizobium* TAL169 which maintained their viability in abiotic stress were investigated, revealed that bacteria replaced their LMWS with others saccharide species, in saline stress (100-400 mM NaCl). When stress conditions was imposed by PEG6000 16% P/V, it was not observed change in LMWS profile (Figure 2). Dardanelli et al (1997) showed that LMWS species was also modified, it was decreased by PEG6000 16%P/V in the growth medium, but it was augmented by high temperature in the strains *Rhizobium* sp. ATCC10317.

A

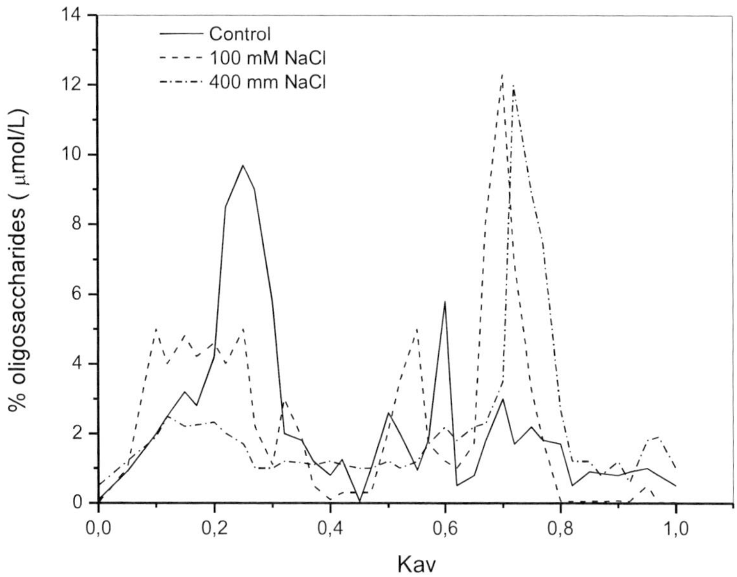

B

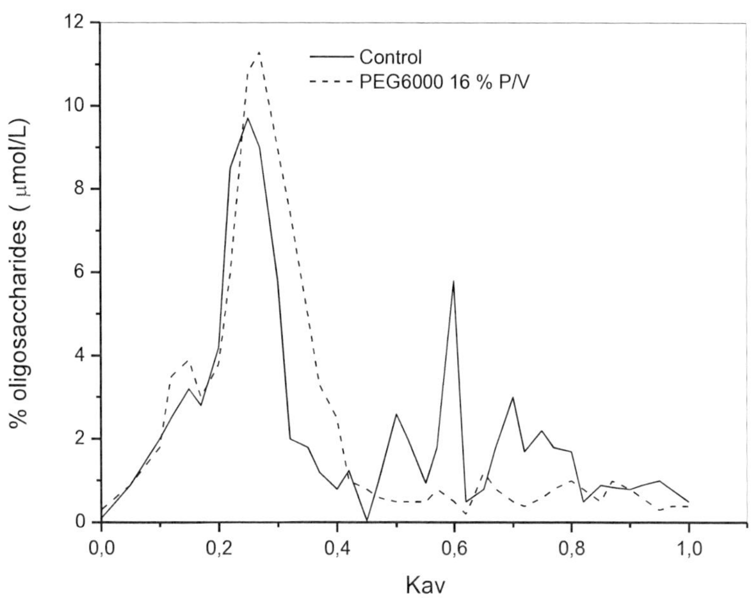

Figure 2. Chromatography on BioGel P4 of periplasmic low molecular weigth saccahrides (LMWS) from stationary phase *Bradyrhizobium* TAL169. **A** control condition (solid line), 100 mM NaCl (dash line) and 400 mM NaCl (dash dot line). **B** control condition (solid line) and PEG6000 16%P/V (dash line). Values represent means of two independent experiments.

Osmotic regulation of *E. meliloti* and *Agrobacterium tumefaciens* cyclic 1,2-ß-glucan synthesis is similar to that of membrane-derived oligosaccharides (MDOs) in *Escherichia coli* (Breedveld and Miller, 1994). High osmolarity was found to inhibit the *in vivo* accumulation of cellular 1,2-ß-glucans in *Agrobacterium* and *Rhizobium* or MDO in *E. coli* (Breedveld and Miller, 1994). Exceptions have been reported with respect to osmoregulation. In several strains of *R. leguminosarum* and of *E. meliloti* strain GR4, the levels of cell-associated cyclic 1,2-ß-glucans were independent of osmolarity of the growth medium (Breedveld et al., 1991). The amount of periplasmic glucans synthesized by *Bradyrhizobium japonicum* decreased when cells were grown in high osmolarity media (Miller and Wood, 1996).

Conclusion

As far as we know, no reports in the literature are available on the adjustment of peanut rhizobia to environmental stress. Our results demonstrate some similar in the response of peanut rhizobia strains. As pointed out by Smith et al. (1988), rhizobia would be expected to have a sophisticated array of mechanisms of adjustment to harsh environmental conditions. Based on our observations, it is tempting to speculate that trehalose and phospholipids may be involved in some of those regulatory mechanisms. Therefore, in our laboratory, experiments are in progress to study the effect of stressful conditions on trehalose and phospholipids metabolisms in peanut rhizobia.

References

Boumahdi, M., Mary, P. & Hornez, J. P. (1999). Influence of growth phases and desiccation on the degrees of unsaturation of fatty acids and the survival rates of rhizobia. *Journal of Applied Microbiology*, *87*, 611-619.

Breedvel, M. W. & Miller, K. J. (1994). Cyclic *β*-glucans of members of family *Rhizobiaceae*. *Microbiology Review*, *58*, 145-161.

Breedveld, M. W., Zevenhuizen, L. P. T. M. & Zehnder, A. J. B. (1991). Osmotically-regulated trehalose accumulation and cyclic beta-(1,2)-glucan excreted by *Rhizobium leguminosarum* bv. *trifolii* TA-1. *Achieves of Microbiology*, *156*, 501-506.

Busse, M. D. & Bottomly, P. J. (1989). Growth and nodulation responses of *Rhizobium meliloti* to water stress induced by permeating and nonpermeating solutes. *Applied and Environmental Microbiology*, *55*, 2431-2436.

Carpenter, J. F. & Crowe, J. H. (1988). Modes of stabilization of a protein by organic solutes during desiccation. *Cryobiology*, *25*, 459-470.

Carpenter, J. F., Crowe, L. M., Crowe, J. H. (1987). Stabilization of phosphofructokinase with sugars during freeze-drying: characterization of enhanced protection in the presence of divalent cations. *Biochimic and Biophysics Acta*, *923*, 109-115.

Crowe, J. H., Crowe, L. M., Carpenter, F. J. & Aurell Winstrom, C. (1987). Stabilization of dry phospholipid bilayers and proteins by sugars. *Biochemist Journal*, *242*, 1-10.

Correa, O., Rivas, E. & Barneix, A. (1999). Cellular envelopes and tolerance to acid pH in *Mesorhizobium loti*. *Current Microbiology*, *38*, 329-334.

Csonka, L. N. (1989). Physiological and genetic response of bacteria to osmotic stress. *Microbiology Review*, *53*, 121-147.

Csonka, L. N. & Hanson, A. D. (1991). Prokaryotic osmoregulation: genetics and physiology. *Annual Review of Plant Physiology*, *45*, 569-606.

Dardanelli, M., González, P., Bueno, M. & Ghittoni, N. (2000). Synthesis, accumulation and hydrolysis of trehalose during growth of peanut rhizobia in hyperosmotic media. *Journal of Basic Microbiology*, *40*, 149-156.

Dardanelli, M. S., González, P. S., Medeot, D. B., Paulucci, N. S., Bueno, M. A. & Garcia, M. B. (2008). Effects of peanut-rhizobia on the growth and symbiotic performance of *Arachis hypogaea* under abiotic stress. *Symbiosis*, *47*, 175-180.

Dardanelli, M., Woelke, M., González, P., Bueno, M. & Ghittoni, N. (1997). The effects of nonionic hyperosmolarity and of high temperature on cell-associated low molecular weight saccharides from two rhizobia strains. *Symbiosis*, *23*, 73-84.

DiRusso, C. C. & Nyström, T. (1998). The fats of *Escherichia coli* during infancy and old age: regulation by global regulators, alarmones and lipid intermediates. *Molecular Microbiology*, *27*, 1-8.

Drouin, P., Prévost, D. & Antoun, H. (2000). Physiological adaptation to low temperature of strains of *Rhizobium leguminosarum* bv. viciae associated with *Lathyrus* spp. *FEMS Microbiology Ecology*, *32*, 111-120.

El-Sheikh, E. A. E. & Wood, M. (1990). Salt effects on survival and multiplication of chick pea and soybean rhizobia. *Soil Biology and Biochemistry*, *22*, 343-347.

Ghittoni, N. E. & Bueno, M. A. (1995). Peanut rhizobia under salt stress: role of trehalose accumulation in strain ATCC 514466. *Canadian Journal of Microbiology*, *41*, 1021-1030.

Halverson, L. J. & Firestone M. K. (2000). Differential effects of permeating and nonpermeating solutes on the fatty acid composition of *Pseudomonas putida*. *Applied and Environmental Microbiology*, *66*, 2414-2421.

Fernández, E. M. & Giayetto, O. (2006). El cultivo de maní en Córdoba. Universidad *Nacional de Río Cuarto*, Río Cuarto.

Kulkarni, S. & Nautiyal, C. S. (2000). Effects of salt and pH stress on temperature-tolerant *Rhizobium* sp. NBRI330 nodulating *Prosopis juliflora*. *Current Microbiology*, *40*, 221-226.

McIntyre, H. J., Davies, H., Hore, T. A., Miller, S. H., Dufour, J. P. & Ronson, C. W. (2008). Trehalose biosynthesis in *Rhizobium leguminosarum* bv. trifolii and its role in desiccation tolerance. *Applied and Environmental Microbiology*, *73*, 3984-3992.

Medeot, D. B., Bueno, M. A., Dardanelli, M. S. & García de Lema, M. (2007). Adaptational changes in Lipids of *Bradyrhizobium* SEMIA 6144 nodulating peanut as a response to growth temperature and salinity. *Current Microbiology*, *54*, 31-35.

Miller, K. J. & Wood, J. M. (1996). Osmoadaptation by rhizosphere bacteria. *Annual Review of Microbiology*, *50*, 101-136.

Mohammad, R. M., Akhavan-Kharazian, M., Campbell, W. F. & Rumbaugh, M. D. (1991). Identification of salt-and drought-tolerant *Rhizobium meliloti* L. strains. *Plant and Soil*, *134*, 271-276.

Potts, M. (1994). Dessication tolerance of prokaryotes. *Microbiology Review*, *58*, 755-805.

Russell, N. J., Evans, R. I., ter Steeg, P. F., Hellemons, J., Verheul, A. & Abee, T. (1995). Membranes as a target for stress adaptation. *International Journal of Food Microbiology*, *28*, 255-261.

Shamseldin, A. & Werner, D. (2005). High salt and high pH tolerance of new isolated *Rhizobium etli* strains from Egyptian soils. *Current Microbiology*, *50*, 11-16.

Smith, L. T., Pocard, J. A., Bernard, T. & LeRudelier, D. (1988). Osmotic control of glycine betaine biosynthesis and degradation in *Rhizobium meliloti*. *Journal of Bacteriology*, *170*, 3142-3149.

Romantsov, T., Stalker, S., Culham, D. E. & Wood, J. E. (2008). Cardiolipin Controls the Osmotic Stress Response and the Subcellular Location of Transporter ProP in *Escherichia coli.* *Journal of Biological Chemistry*, *283*, 12314-12323.

Suutari, M. & Laakso, S. (1994). Microbial fatty acid and thermal adaptation. *Critical Review Microbiology*, *20*, 285-328.

Welsh, D. T. (2000). Ecological significance of compatible solute accumulation by microorganism: from single cells to global climate. *FEMS Microbiology Review*, *24*, 263-290.

Wood, J. M., Bremer, E., Csonka, L. N., Kraemer, R., Poolman, B., van der Heide, T. & Smith, L. T. (2001). *Comparative Biochemistry and Physiology Part A*, *130*, 437-460.

Zahran, H. H. (1999). *Rhizobium*-legume symbiosis and nitrogen fixation under severe conditions and in arid climate. *Microbiology and Molecular Biology Review*, *63*, 968-989.

Zorreguieta, A., Cavaignac, S., Geremia, R. A. & Ugalde, R. A. (1990). Osmotic regulation of 1,2-*b*-glucan synthesis in members of the family *Rhizobiaceae*. *Journal of Bacteriology*, *172*, 4701-4704.

In: Organic Farming and Peanut Crops ISBN: 978-1-60876-187-6
Editors: D. C. Grossman, T. L. Barrios, pp. 163-189

Chapter 7

The Paradigmatic Nature of Organic Farming: Evidence from Agricultural Professionals' Attitudes Towards Organic Farming and Genetic Engineering

Sarah Wheeler
Centre for Regulation and Market Analysis, School of Commerce
University of South Australia, Adelaide, SA 5000

Abstract

This paper discusses the concept of scientific paradigms and draws parallels with paradigms in agriculture, and in particular focuses upon organic farming and genetic engineering. There is anecdotal evidence that they are both potential new agricultural paradigms. Different writers have referred to both as new 'paradigms' in agriculture, though the qualitative evidence is stronger for organic farming than genetic engineering. The attitudes of agricultural professionals are important as they are responsible for creating, developing and diffusing innovations as well as informing and educating farmers and the public. This paper summarizes past research on the attitudes of Australian agricultural professionals (n=185) towards organic farming (Wheeler 2008) and genetic engineering (Wheeler pending). These two sets of results are compared to assess if professional attitudes to the two innovations are driven by similar influences or are quantitatively significantly different. In addition, analysis using bivariate ordered probit regression is conducted in this study on the nature of professionals' answers to the net benefits of the two innovations themselves. Do professionals tend to be negative or positive about both innovations, or positive about one and negative about the other? From a comparison of the results, it is shown there is some quantitative evidence that organic farming represents a potential new sub-paradigm in agriculture, but only very weak evidence that genetic engineering is a potential new paradigm.

Introduction

Most scientists believe that science operates via the rationalistic 'scientific method', which postulates that science is pure, and that propositions and methods are arrived at only by efficient, logico-empirical operations. Personal, political and power influences play no role. However, there is considerable evidence to suggest that science does not always operate in a manner consistent with this methodology.[1] The question is, why? Part of the reason includes professionals' individual psychological factors and belief systems. In particular, the existence of paradigms may help explain why the rationalistic scientific method does not always prevail in science.

In 1962, Thomas Kuhn wrote what is claimed to be his classic work: *The Structure of Scientific Revolutions,* in which he introduced the term 'paradigm'. Kuhn used the term to explain the nature of scientific debates. A paradigm was a dominant theory in science, or 'normal science'. The term is still used today to convey the implicit assumptions underlying and guiding scientific inquiry, as well as describing the fundamental currents of social change. In Kuhn's view, science evolves in the following way:

Pre-science (pre-paradigm) → normal science (paradigm) → crisis → revolution → new normal science → new crisis (and so on).....

In the pre-science state, a number of schools of thought compete against each other for dominance. The scientific field becomes more structured and directed when one school of thought overshadows the others, and becomes the dominant paradigm or 'normal science'. The framework helps define the nature of the world and a way of looking at things: it is a set of shared assumptions, beliefs, dogmas, conventions and theories. Working within a paradigm implies rules and boundaries, which limits what questions can be asked, how research is conducted, and the type of answers that are found. Kuhn argues that scientists are not bold adventurers discovering new truths; rather they are puzzle-solvers working within an established view of the world. The normal state of science is said to be dominated by one paradigm, although it was later agreed by Kuhn that a number of sub-paradigms can also exist at one point in time. Lakatos (1970) agrees that a number of research programmes can exist at the same time, and his research programmes closely resemble Kuhn's paradigms. Some writers prefer Lakatos' description of research programmes to Kuhn's paradigms as being more descriptive of scientific history, especially in the field of economics (i.e. Blaug 1975).

Paradigms can be a useful framework in science. They can eliminate the need to continually question basic assumptions or issues and therefore reduce wasted time. However, the existence of dominant theories and power structures in science may result in the suppression of undesirable opinions, unoriginality in research and emphasis on trivial topics, a denigration of non-specialists scientific knowledge, and an increase in fraud and favouritism in hiring and promotion (Martin 1999, Manwell & Baker 1988, Broad & Wade 1982, Martin 1979, de Grazia 1966). Another potential consequence of paradigms in science is path-

[1] Bauer (1984), Burr (2002) and Krimsky (2000) provide histories of a number of innovative scientists whose ideas were rejected outright by mainstream science. Other deviations include: suspicions attached to theories which originate in particular countries; less rigorous checks on scientists of long standing and demonstrated ability; the withholding of theories until the release impact can be maximised; profit-orientated research goals; and reluctance to let go of particular theories no matter what the evidence suggests.

dependence, where one research area becomes superior to another simply because of the research funds spent and its future use and development. However, had the other new theory had similar levels of funds, use and development, it would have been socially superior. The implications are that socially beneficial research and innovations may be lost. The path-dependence argument has been strongly criticised (i.e., Liebowitz & Margolis 1994 and 1990), yet it remains as a potential negative consequence of the existence of paradigms.

A new paradigm often attracts controversy and criticisms when the intellectual attraction of the research network is exhausted, or when external conditions lead to questioning the application of the current paradigm. A 'paradigm shift' is a change in the dominant theory of science. One of the questions Kuhn seeks to answer is why individual scientists change their allegiance from one paradigm to the other (or why they work in a particular field). Kuhn suggests that such a change is not purely rational, that other factors come into play, as scientists have different values, standards, goals and principles. In one way, opponents and proponents of different paradigms speak different languages. Each group cannot *compel* the other to accept their view, but they can attempt to *persuade* each other to change views. There will be an increasing shift in professional allegiance as more and more scientists are converted to a particular view. In Kuhn terms, the "revolution" occurs when the majority of scientists have shifted to a particular stance. This may happen quite quickly, but at other times it could take centuries for the shift to occur.

There is no one absolute test to show that a new theory is becoming the new paradigm. A variety of factors will influence how scientists view new paradigms and either maintain or switch allegiance. This paper assembles a number of economic, sociological, philosophical and psychological theories on how scientists react to new ideas and then compares quantitative analysis on two agricultural innovations to indicate the presence of a new paradigm. Previous research in this area has primarily been current or historical case studies and generally qualitative in nature (for example, Barnes *et al.* 1996, Feyerabend 1988, Latour and Woolgar 1986, Latour 1987, Merton 1973).

ORGANIC FARMING AND GENETIC ENGINEERING

Genetic engineering and organic farming represent two key innovations in agriculture that are both lauded and deplored. Different parties represent the two innovations as radical alternative ways to address current sustainability crises in agriculture (i.e., National Academy of Sciences 2000, Pretty 1995). Like Wynen (1996), we draw an analogy of a paradigm shift in science with a paradigm shift in agriculture, although they are strictly not the same thing.[2] But can they be described as potential new paradigms in agriculture? That is, will the current system of conventional agriculture give way to a new system that looks more like organic farming, or a new system of biotechnologically-engineered research and products? Or are both systems simply comparable to the current system?

[2] Agriculture is not strictly a scientific field; it draws on knowledge and practices from many different scientific fields. It is better described as a management system, with conventional agriculture representing the dominant management system today.

Organic Farming

Organic farming is being increasingly adopted around the world and it is becoming more mainstream. The goal of organic farming is to give priority to long-term ecological health, such as biodiversity and soil quality, rather than short-term productivity gains. A number of studies have provided scientific evidence on the environmental, economic and health benefits of organic farming (Macilwain 2004, Shepherd *et al.* 2003, Nelson *et al.* 2004, Mäder *et al.* 2002, Stolze *et al.* 2000, O'Riordan & Cobb 2001). However, many agricultural professionals still question its financial viability, environmental credentials and overall efficiency and productivity, and reject it as an alternative farming system (e.g., Trewavas 2001, Avery 1995). Organic farming in Australia receives extremely little government research or extension support (Wynen 2003), far less than its current adoption by farmers.

Organics represents a way of farming – but it can also represent a way of scientific research.[3] Wynen (1996) argued there was strong qualitative evidence of a paradigm shift in agriculture, from the dominant paradigm of conventional agriculture towards organic farming. This qualitative evidence included highlighting problems within the dominant paradigm (such as the crises within conventional farming, along with attempts to 'patch-up' these problems); increasing argument between agricultural professionals and farmers that is unable to be resolved (there is growing polarisation of opinions between conventional and organic proponents which indicates the incommensurability of their views); and the increasing popularisation of the potential new paradigm (popularisation occurred through increasing organic journal and book publications, increased tertiary education, development of extension channels and farmer financial support, and increasing farmer adoption of organic farming). Such indications led Wynen (1996) to suggest that organic farming does represent a paradigm shift in agriculture.

Bues and Dunlap (1990) also sought to find whether conventional agriculture and organic farming represent two different paradigms (but they did not use the exact same definition as Kuhn). They concluded (p. 597): "in short, all indications are that the contemporary debate in agriculture seems to have the makings of a true paradigmatic conflict". Other writers are increasingly referring to organic farming as a potential "new paradigm" in agriculture (Dimara *et al.* 2003, Abaidoo & Dickinson 2002, Allen 1996, Dahlberg 1986, Merrill 1983). There is enough qualitative evidence to suggest that organic farming may represent a potential new paradigm – both as a system of farming and as a new form of agricultural research. Later in this chapter we test how much quantitative evidence exists for organic farming to be a paradigm shift in agriculture.

---

[3] Conducting organic agricultural scientific research is difficult for scientists due to the need to consider the whole farm and its ecological system, not just parts of the system. There are five main ways organic research differs from conventional: (1) it applies consistently multidisciplinary approaches to solutions; (2) it integrates stakeholders into research; (3) it rescales continually the focus of research work in the same project; (4) it considers and decides on long term impacts; and (5) it considers individual, social, ethical and philosophical frameworks of agriculture and food in addition to natural science (Niggli 2002).

Genetic Engineering

Genetic engineering (GE) is the modern innovative form of biotechnology, known also as recombinant-DNA technology. Some of the major benefits of genetic engineering that are cited include agronomic (increased yields, reduced costs), health (increased vitamins) and environment (reduced pesiticides, less land needed for production) (King 2004 and 2003, National Academy of Sciences 2000, Beyer *et al.* 2002). Critics argue that its benefits are uncertain and often exaggerated, while potential problems include health, environment, trade and legal issues (Independent Science Panel 2003, Quist & Chapela 2001, Ewens & Pusztai 1999). Australia invests strongly in genetic engineering, with public sector related R&D biotechnology investment in agricultural, vet and environment representing 35% of total R&D expenditure in these fields in 2002-03 (Hopper & Thorburn 2005).

Some scientific writers have described genetic engineering as a new paradigm that is going to change the face of science (Leisenger 2000, ISSC 1993, Miller 1993), but we do not know of any writers in sociology who have applied the term. On the one hand, scientists argue that biotechnology basically involves the same modifications that have been taking place for thousands of years, with the modern form of genetic engineering adding additional information (Borlaug 2000, Prakash 2001). Critics of genetic engineering believe that rDNA techniques are fundamentally different from traditional technology and should be evaluated differently (ISP 2003, Gepts 2002), and that it will not solve the current problems inherent in conventional agriculture today. Research on genetic engineering does not seem to have been restricted or held back, as in the beginning there were no scientific restrictions imposed on molecular biological scientists, and although concerns began to be raised by other scientists in the late 1960s, no direct regulation or public involvement resulted. In addition, funding and research support were never limited. Government involvement and scientific criticism increased in the 1980s but it was not until the 1990s that the public (and other scientists) really became involved in the debate and restrictions started being imposed.

There is some evidence for and against genetic engineering, causing a paradigm shift in agricultural scientific research. Evidence for genetic engineering includes public and some scientist resistance, the ground-breaking developments in scientific research, and the framework around which genetic engineering is considered. Evidence against genetic engineering includes the history of the technology, the ready availability of research funds and the tacit acceptance of the techniques by the majority of scientists. The overall balance of evidence may imply that genetic engineering is an important new innovation/technology only, but not a paradigm shift, at least not in Kuhn's sense of a 'revolution'. Some writers argue that there are indications that it has been absorbed into the current dominant practice of agricultural scientific research and is fast becoming part of that paradigm (Egri 1994). Whether this is the case or not is explored in this paper.

PREDICTING A PARADIGM CHANGE

The current conflict in agriculture between organic farming and conventional agriculture; between genetic engineering and conventional agriculture; and between genetic engineering and organic farming exists at both societal and scientific levels of a debate. At a broad level, the debate is fundamentally philosophical and value-laden, which is hard to resolve with empirical evidence. Indeed, it is impossible to separate out agricultural practices from the beliefs and values that underlie them. The development and acceptance of these innovations has not been a purely rational process, but one that has involved conflicting beliefs, values, perceptions and social interaction.

The following is an attempt to assemble, discuss and state various hypotheses on how scientists react to new ideas, using ideas from philosophers, historians of science and sociologists about the reception of radical changes in scientific beliefs. It is discussed why some scientists may be encouraged to accept the current paradigm and why others consider alternative theories. This paper develops a number of hypotheses for agriculture which are then compared to data for organic farming and genetic engineering. The hypotheses are not by any means considered to be definitive or final; they are open for consideration with the aim to generate further debate, discussion and research.

Status Effects

One of the most cited influences on scientists' acceptance of new paradigms relates to status effects (i.e. Dolby 1979, Merton 1973, Mulkay 1972, Kuhn 1970, McClelland 1969). Philosophers of science suggest that science control is maintained through socialisation; through the exercise of authority; and by means of social exchange. Innovations are said to come predominantly from those who are new to a particular area or new to research altogether and who have not yet acquired a strong commitment to the current paradigm. Nonconformity will therefore generally occur within academic status classes, namely the young and those of high status. Members of the middle-status category are least likely to deviate from group opinion or group norms, while persons of low (either because they are young or new to the area at hand) or high (either those who have been around for a while, are a leader or professor or journal editor, etc.) status are much more likely to be nonconformist. Those of high status obtain decreasing returns from undertaking further research within the current paradigm and are less likely to feel stigma from their colleagues about deviating from current thinking. However, the effect is not certain because those with high status still have a lot to lose in deviating from different research patterns. The young gain little from producing highly predictable results, while they potentially gain a lot from innovating. Kuhn (1970; 151) describes what is known as the "Planck Effect":

> "Max Planck ..... sadly remarked that 'a new scientific truth does not triumph by convincing its opponents and making them see the light, but rather because its opponents eventually die, and a new generation grows up that is familiar with it'".

Evidence such as the age of some famous scientists (Darwin was 29 when he first wrote about natural selection and Einstein was 26 when he published his theories) supports the theory that being younger is associated with revolution. However, Kuhn believed this generalisation badly needed systematic investigation. Popper (1970) suggests from his personal experience that it is young science students that are more likely to be non-revolutionary and accept the ruling dogma of the day – hence age is not a good criterion. Indeed, professionals who are about to retire may be more likely to want to consider alternatives as they no longer feel the pressure associated with self-interest. Merton (1973) found that the evidence on age differentials in receptivity to new ideas in science was thin and uncertain. One of the conflicting corollaries is that age is also associated with other measures of status and specialisation, hence once these effects are removed, being older may be associated with championing new paradigms. Although it is recognised that the effect of age on acceptance of new paradigms is far from certain, because most of the literature has supported the generalised link between being young and accepting alternative paradigms, this is the hypothesis that is tested here.

Finally, many scientists swap fields during their career to work in another scientific area (one that they have not been trained in). Personal reflections of such scientists suggest that they believe they are able to see a broad picture within that field, without being hampered by 'knowing what everyone else knew'. Newcomers have often introduced radical new ideas into their adopted scientific field (Merton 1973). Thus:

H1: Younger professionals are more likely to be positive towards an alternative paradigm than older professionals.
H2: Professionals who have worked longer in their area are less likely to be positive towards an alternative paradigm than professionals who have worked less in their area.
H3: Professionals who are of 'high-status' (classified as a leader in their job) are more likely to be positive towards an alternative paradigm than non-leaders.

Self-Interest

Economic theory suggests that professionals will be influenced by their own self-interest. Self-interest can lie in the salary obtained under the current agricultural system, or the considerable costs associated with tertiary education, or in the funding available to current research (i.e., Barnes *et al.* 1996, Barnes and MacKenzie 1979, Martin 1979, Mulkay 1979). Generally there is pressure on scientists to do certain types of research, which is predisposed towards the types of topics that can be researched and the economic and non-monetary rewards that are available from pursuing certain paths. Kuhn suggests that scientists who have more to lose from abandoning the current paradigm will not support alternative paradigms. It is only rational that professionals can be expected to look after their own best interests, and the more self-interest (via, say, salary or work relevance) professionals have in the research area of the current paradigm, the less likely they will support an alternative.

Scientists have undergone prolonged training to accept and repeat the values, standards and assumptions of science as presented by teachers and the education system. Therefore, there may be a vested interest associated with increased tertiary education and acceptance of new paradigms, with scientists who have incurred large education costs from the current system more likely to reject alternative paradigms. Francis Bacon was the first to suggest that increased specialised education may introduce bias and preconceptions (Riggs 1992). On the other hand, advanced education may be a variable that allows for a more questioning mind. Research on people's consciences and obedience to scientific authority indicates that individuals with advanced education may be more likely to hold on to the perception that they are a legitimate decision-making authority and will therefore question decisions made by people who are meant to be authorities (Milgram 1983). Also, education is a sunk cost, hence should not affect scientists' current or future behavior. Having increased education may mean a professional is more positive towards an alternative paradigm; nevertheless, it is the previous proposition that is tested due to the predominance of writers suggesting this link.

Scientists who are advocates or tacit supporters of a new paradigm believe that they are 'right' because they have higher knowledge about the area. Those who reject an alternative paradigm outright may feel that they do not need to know about it because its underlying principles show that it is clearly wrong. Therefore, increased knowledge of a paradigm increases the likelihood that professionals think more favorably about it. At the same time, critical skeptics of alternative paradigms may argue that scientific knowledge has not been correctly interpreted by the advocates or tacit supporters, and therefore critics rate their knowledge higher and proponents' knowledge deficient, which may weaken the above hypothesis.

H4: Professionals who have a higher salary under the existing scientific structure are more likely to be negative towards alternative paradigms.

H5: Professionals who have a higher percentage of research relevance tied to an alternative paradigm are more likely to be positive towards the alternative.

H6: Professionals who have more tertiary education are more likely to be negative towards an alternative paradigm than professionals with less tertiary education.

H7: Professionals who have a higher knowledge of an alternative paradigm are more likely to be positive towards that paradigm than professionals with less knowledge.

Specialisation

The professionalisation process for individuals has been described by sociologists and others as a process of acculturation that occurs in often subtle ways, resulting in a state of mind that is characteristic of a member of a particular profession or occupational group. Becker and Carper (1956) identified four elements that individuals use to weave occupational identity, which are still used today as a basis for defining the professionalisation process. These elements are (a) occupational title, and associated ideology; (b) commitment to task; (c) commitment to particular organisations or institutional positions; and (d) significance for one's position in the larger society. One's self image or the self image of one's profession will therefore affect one's conduct, attitudes and perspective. Occupational title and organization employment are tested here. Merton (1973) also discusses the importance of the roles played within science. Different professions (such as scientists, researchers, academics and extension officers) within science play different roles, and are employed in organizations such as government, industrial scientific organisations and universities. Popper (1970) distinguishes between 'applied' and 'pure' scientists -- applied scientists are those who are more content to live within 'normal science' and 'pure' scientists are not. Popper argued that pure scientists are comparatively rare within science, hence research scientists are on average expected to be more applied than pure. It is expected that given the nature of their job, extension officers are more applied scientists than pure scientists. Conversely, it is hypothesized that professionals employed within universities are more likely to be pure scientists.

Finally, specialisation within a profession may reinforce the structure and worldview inherent in the selection and organisation of scientific knowledge because of the reduction of contact with wider political and social issues. This may increase the potential loss in personal research from a change in paradigm (Martin 1979, Popper 1970), leading to an immense restriction on the scientist's vision and considerable resistance to paradigm change. Therefore, professionals that work across many areas of agriculture (in, say, fields such as natural resource management) are more likely to be positive towards alternatives. Thus:

H8: Scientists are more likely to be negative towards an alternative paradigm (such a hypothesis assumes that generally there are more 'applied' than 'pure' scientists).
H9: Extension officers are more likely to be negative towards an alternative paradigm.
H10: The existence of any occupational effects for organisation of employment indicates professionals are more likely to be negative towards an alternative paradigm.
H11: Professionals whose job is less specialised (i.e.: they work in the field of natural resource management) are more likely to be positive towards an alternative paradigm.

Gender

Feminist epistemology and the feminist critique of science and technology literature suggest that women and men are fundamentally different when it comes to holding views about research, science and technology. Such literature (reviewed in Buttel and Goldberger 2002; 27) argues that females tend to be "more critical, more questioning and more public-regarding", and hence would be less likely to be captured by the current paradigm. On the other hand, there is an argument in the literature that females are more financially dependent upon their jobs and hence would be less likely to support alternative paradigms where the payoffs are risky. Nevertheless, it is the first proposition that is tested.

H12: Female professionals are more likely to be positive towards an alternative paradigm.

Principles

Agricultural professionals are taught to be useful, generally because they are scientists who are responsive to the demands of specific users. Things that are considered to be 'uneconomic' are often considered not worth pursuing and are not the prime goal of research (Hadwiger 1982; 56). A scientist who is preoccupied with trying to be useful may not notice wider social implications. It is suggested that most get immediate satisfaction out of the belief that their research has made feeding the world's population possible. Krimsky (2000) suggests that a scientist who is traditional, positivist, private and sponsored by government or industry will tend not to become involved in advocating or supporting environmental health hypotheses.

Empirical evidence seems to support Krimsky's above argument. A survey of 70 Australian environmental scientists found that most scientists provided policy advice in the form of apolitical advisory roles rather than active participation. Over half of them (54%) believed that their career prospects or research funding would be jeopardised by speaking out on environmental issues (29% were unsure), while 11% had been directly disadvantaged because of their environmental views (Wilson & Barnes 1995). It is suggested that agricultural environmental views would not be voiced as strongly as agricultural production views.

H13: Professionals are more likely to be positive towards an alternative paradigm if it is seen to be associated with profitability. Their views on such issues will be more important than any other potential benefits.

H14: Professionals who believe conventional agriculture is unsustainable are more likely to be positive towards an alternative paradigm. Such views will not be as strong in influencing overall attitudes as their views on production or profitability.

Kuhn suggests scientists who support alternative paradigms are more involved in crises (problems surrounding the current paradigm) and are more likely to be vocal in advocating new positions. Krimsky (2000) suggests controversies in science arise when facts are uncertain, values are in dispute, stakes are high and decisions are urgent. There are no clear scientific ethics of engagement pertaining to these sorts of controversies. Ethical problems include the release of uninterpreted data, conjecture sensitivity, prepublication release of research, and conjectural risk. Responses of scientists to controversial hypotheses can be classified into four types: (a) advocacy scientists who feel it is their responsibility to support and advocate controversial hypotheses, even if the evidence has not been absolute. They call for regulation of and action against questionable products; (b) tacit supporters who believe that the controversial hypothesis is likely to be true, and therefore rethink their own research; (c) critical skeptics whose arguments lead to the strengthening, weakening or falsification of a controversial hypothesis. Critics support acting only when the evidence is undeniable and the science is 'sound'; and (d) silent skeptics who will not believe that there is enough evidence supporting the controversial hypothesis, and refuse to consider its potential effects (Krimsky 2000). Scott, Richards and Martin (1990) show that when scientists become involved with non-mainstream theories they often become participants in the "fierce fight to construct reality", and their position generally changes from that of neutral, symmetrical analysis to active intervention in the dispute. An outspoken critic of pesticides, the entomologist Van den Bosch (1978; 48-49), states that he understands why many scientists are not willing to participate in public debate on technological issues and that the lack of fair debate over the issue forced him to also abandon reasoned debate.

> I have become increasingly involved in the roaring pesticide controversy – a vicious, nerve-wracking imbroglio that has turned my entomological niche into a veritable hornet's nest. What is most saddening is that, as I have become increasingly enmeshed in the pesticide hassle, I have turned into a ruthless gut-fighter in a slugfest without rules or a semblance of fair play. This metamorphosis has been forced upon me because it is the only way that I can hold my own in the shoot-out, or for that matter even survive it. It is difficult for a scientist to play this way. We are largely a tribe of preoccupied people who just want to be let alone to do our thing. Since we lack the appetite for slashing combat, or the skills to survive connivance, we usually avoid confrontation.

Therefore, those scientists who are advocating stronger involvement from government are more likely to be favourable towards alternative paradigms, because scientists within mainstream science are unlikely to actively intervene.

There is currently strong debate over the appropriateness of the increasing use of intellectual property rights and patenting in agriculture and the influence of private funds on research outcomes (Huffman & Tegene 2002, David 2000, Mazzoleni & Nelson 1998, Drahos 1996). Tisdell (1991) suggests that patented inventions and intellectual property rights favour knowledge associated with conventional agriculture, hence private research is distorted. He believes a follow-on effect also often occurs with publicly funded or conducted agricultural research. Huffman (2001) argues that major problems are arising with overlapping patents, excessive breadth, lockouts and issues with freedom to operate. Universities cannot be 'open' if their funding or research agenda has been captured by a number of large companies. Professionals who are critical of certain research influences in the current agricultural paradigm may be more likely to be favourable to alternative paradigms where such issues may not exist.

H15: Professionals who advocate the need for a more pro-active government policy role for the alternative paradigm are more likely to be positive towards it.

H16: Professionals who view the increasing role of intellectual property rights and patenting in agricultural research positively are more likely to be negative towards an alternative.

H17: Professionals who think that private funding in agricultural research does not influence research outcomes are more likely to be negative towards an alternative paradigm.

Information Choices

Due to the quantity and quality of information choices available, professionals' code (i.e., rate or weight) messages received. Agricultural information choices include those from scientific sources (such as conferences, journals, peers, scientific books, university) and other sources (media, farmers, industry, books). Different information channels have differing weights (Allen 1969). Firms and organisations control the intake of information to avoid being overwhelmed by setting up a 'system boundary' to provide barriers between information on the 'inside and on the outside' (Price 1969; 151).

Another form of setting system boundaries is how professionals view the involvement of the public, media and other lay groups in decision making, as compared to recognised scientific sources. One of the main tenants of the rationalistic scientific method is that scientists are the best interpreters of scientific data, and results should only be communicated to the public after it has gone through a rigorous peer review system. However, if scientists do not believe that their concerns have been given a fair consideration within science (for reasons such as those provided in footnote 1), then those scientists often turn to the public for a hearing of their beliefs (Burr 2002, Krimsky 2000, Scott, Richards & Martin 1990). The involvement of the public and other non-scientific sources can be viewed positively or negatively by scientists. If they view the involvement of lay groups as a main barrier against the diffusion or employment of scientific methods, then they are likely to be more favorable to the current paradigm. Thus:

H18: Professionals that gain information primarily from 'inside sources' (primarily scientific sources) are more likely to be negative towards an alternative paradigm.

H19: Professionals who name public attitudes, lobby groups and media influences as the main barrier against the diffusion of innovations are more likely to be negative towards an alternative paradigm.

Overall Changes in Views

Finally, for a paradigm shift to occur in agriculture, professional views towards the innovations must be changing, with more professionals 'converting' to the view. Kuhn suggests that a paradigm change is decided by the number of scientists who advocate the reigning paradigm as against those who advocate the alternative. It is important to note that this hypothesis is not a test of how scientists react to new paradigms per se, but more of an indication of the possibility of a paradigm shift occurring.

H20: More professionals must have positively changed their view towards the alternative paradigm than those that negatively changed their views over the past few years.

There are a number of sociological and philosophical hypotheses that could not be tested within this paper. Some of the more important remaining hypotheses that could be explored in future research include examining the 'gatekeeper effect'. If professionals view themselves as 'gatekeepers of science', (i.e., they may be peer reviewers for a journal, members of an editorial board or members of a scientific group) then the more they may support the current paradigm. Also, the influence of professional goals could not be assessed; that is, how does their stated job goal (i.e., feed the world, scientific curiosity, business development, etc.) affect their views of alternatives? Other factors that have often been mentioned include religious, insecurity issues and political factors. These and other hypothesis tests are left to future research.

Survey Methodology and Model Specification

Acceptance of an agricultural innovation can differ for a number of levels: research, extension, farmer, and community. Agricultural professionals (scientists, extension officers, researchers and academics) play a very important role in creating, diffusing and informing the public about agricultural innovations.[4] Previous papers (Wheeler (2008) and (pending)) have explored the influences on agricultural professionals' beliefs towards organic farming and genetic engineering, respectively. These studies used data from a telephone survey conducted in mid 2004. The sample frame for the survey consisted of two groups: a general and a

[4] An agricultural professional was defined as either providing agricultural advice to farmers; conducting agricultural research; or teaching agriculture at university. The sample was limited to public bodies.

targeted group. The general group of agricultural professionals was limited to State and Commonwealth bodies based in South Australia (SA). Respondents were randomly selected from each group and telephoned, with 119 surveyed.

To supplement the general sample, professionals who had actual experience with organic farming across Australia were specifically sought for additional telephone interviews. The purpose of this supplementation was to obtain enough records of professionals with organic research/extension experience in order to be able to compare their attitudes and knowledge with those who do not have such experience (as only a small percentage of professionals do have such experience).[5] 66 interviews were conducted with professionals around Australia, giving a total sample of 185 respondents, and a total response rate of 96%.[6]

The dependent variables used by Wheeler (2008 and pending) were the respondents' answers to the questions on whether the benefits of each innovation outweighed its costs or risks. Professionals either responded "no," "unsure," or "yes". The answer "no" indicated a belief in no net social benefits associated with either innovation, "unsure" indicated uncertainty due to unknown factors, and "yes" indicated a belief in net benefits. Therefore as it stands, the dependent variable represents a proxy benefit cost ratio. It attempts to assess all the aspects of each innovation and determine whether overall, the innovation is considered worthwhile. Two ordered probit models were used to evaluate factors that influenced overall views towards each innovation. Appendix 1 provides the variable coding and questions asked in the surveys.

Empirical Results

With the exception of hypotheses 15, 19 and 20 (which could not be tested in the regressions due to their inappropriateness as independent variables), all the remaining hypotheses can be tested using the ordered probit results. Table 1 depicts and compares the seven broad groups of independent variables that influence views towards organic farming and genetic engineering. Factors significant in the organic farming model included working age, Commonwealth Scientific and Industrial Research Organization (CSIRO) employment, being part of the targeted sample, organic knowledge, information sources, beliefs in the innovation's environment sustainability, food quality, profitability and innovativeness. Factors significant in the genetic engineering model included age, ethnicity, farm background, scientist job title, extension officer, genetic engineering research relevance, views on the sustainability of conventional agriculture and intellectual property rights and patenting in agriculture, and views on genetic engineering's products profitability, environmental friendliness and long term research needs. Further discussion on the regression results is provided later.

---

[5] A targeted database of over 100 academics, scientists, extension officers and policy makers from 16 Australian public organisations was created. Respondents were randomly selected and phoned if permission from the head of the organization had been received for ethics clearance purposes. If not, individuals were passed over if selected. The sample sizes for the random and targeted groups were both statistically significant at a 7% level of precision and a 95% confidence level.

[6] The response rates of each population were not statistically different from each other, though the targeted population was slightly higher than the random population (97% vs 96%).

Table 1. Determinants of Views towards Organic Farming & Genetic Engineering (ordered probit regression)

	Organic Farming			*Genetic Engineering*		
	Coef.	z	P>|z|	Coef.	z	P>|z|
Age	-0.13	-1.53	0.13	-0.14	-1.84	0.07
Age^2	0.00	1.66	0.10	0.00	1.97	0.05
Working Age	-0.03	-1.69	0.09	-0.01	-0.55	0.58
Ethnicity	0.63	1.19	0.24	1.55	2.18	0.03
Male	-0.19	-0.67	0.50	0.22	0.99	0.32
Having a Farm Background	-0.02	-0.11	0.91	-0.51	-2.65	0.01
Years of Tertiary Education	0.13	2.12	0.03	-0.06	-0.76	0.45
Salary	0.00	-0.10	0.92	0.00	0.72	0.47
Job title is Scientist	0.00	0.00	1.00	0.63	2.63	0.01
Classified as a Leader in their Position	0.18	0.46	0.65	-0.04	-0.12	0.91
Working predominantly as Extension Officer	-0.06	-0.23	0.82	0.40	1.70	0.09
Employed by a state organization	0.02	0.07	0.94	0.16	0.50	0.62
Working in Natural Resource Management	0.35	1.13	0.26	-0.07	-0.26	0.80
Working in Broadacre and/or Grazing	0.04	0.14	0.89	0.12	0.49	0.62
Employed by CSIRO	-0.89	-1.77	0.08	0.13	0.25	0.80
Part of the Targeted Sample	0.65	1.93	0.05	-0.14	-0.59	0.56
Organic/Genetic Engineering Knowledge	0.24	1.76	0.08	-0.05	-0.41	0.68
Naming Media as first main source of Info.	-0.34	-1.27	0.21	0.29	1.15	0.25
Naming Scientific sources as first main source of Info.	-1.16	-3.55	0.00	-0.07	-0.25	0.80
% of Research Time Relevant to Organic Farming/Genetic Engineering	0.01	1.16	0.25	0.01	2.12	0.03
Views on Conventional Agric. Sustainability	0.12	1.10	0.27	-0.24	-2.63	0.01
Views on agric. intellectual property rights & patents	0.08	0.79	0.43	0.23	2.37	0.02
Views on private funding of agric. research	0.02	0.14	0.89	-0.16	-1.53	0.13
Views on Organic Farming's*/Genetic Engineering's Profitability	0.25	1.65	0.10	0.39	2.60	0.01
Views on Organic Farming's Innovativeness/ Beliefs on uncertainty of GE and need for long-term research	-0.37	-2.70	0.01	-0.65	-4.56	0.00
Views on Organic Farming's/Genetic Engineering Food Quality	0.27	2.33	0.02	0.11	0.74	0.46
Views on Organic Farming's/Genetic Engineering Environmental Sustainability	0.70	5.16	0.00	0.51	3.57	0.00
/cut1	-0.64	1.95		-3.39	2.00	
/cut2	0.65	1.96		-2.06	2.01	
Number obs.	185.00			185.00		
Log Likelihood	- 115.44			- 137.29		
Wald Chi2	147.91			129.83		
Prob>chi2	0.00			0.00		
Pseduo R2	0.40			0.32		

Sources: Adapted from Wheeler (2008, pending) and further analysis

Notes: All models calculated with Huber/White heteroskedasticity-consistent standard errors and covariance. Cuts report standard errors.

* In order to make similar comparisons between the two, the variable sign for organic farming's profitability was reversed.

Table 2. Professionals' Views on a More Pro-active Government Policy Stance by Type of Innovation

Views	Organic Farming (General)	Genetic Engineering (General)	Organic Farming (Total Sample)	Genetic Engineering (Total Sample)
Positive	3.76	3.48	4.03	3.41
Sample size (n)	*44*	*54*	*94*	*69*
Unknown	3.26	3.57	3.27	2.66
Sample size (n)	*35*	*29*	*46*	*58*
Negative	2.6	2.14	2.59	1.91
Sample size (n)	*40*	*29*	*47*	*58*

Note: Responses were based on a 5 point Likert scale, from Strongly Disagree (1) to Strongly Agree (5).

The psuedo R-square value describes how well the data fit the model. A significant amount of variation was explained. This prediction result compares favourably with other ordered probits.

In the models in Table 1 above, similar factors are thought to influence both genetic engineering and organic agricultural views. A question may be raised over the degree to which innovation views are inter-related: do professionals tend to be negative or positive about both innovations, or positive about one and negative about the other? This paper employed bivariate ordered probit regression which fits maximum-likelihood two-equation ordered probit models of our two dependent variables (views towards genetic engineering and views towards organic farming) on the independent variables for each model. Results are shown in Appendix 1. It was found that the error terms of the two models were not correlated (by the Wald test of independent equations of if rho is significantly different from zero); hence, believing that genetic engineering has net social benefits has no influence over believing that organic farming has net social benefits. Hence, it is possible to model the two beliefs separately and the results shown in Table 1 can be discussed further with confidence.

Individual Tests of Remaining Hypotheses

Hypothesis 15 suggests professionals who advocate the need for an increased government policy role are more likely to be favourable towards a new paradigm. In the survey, professionals were asked whether they agreed or disagreed with two questions: "Government should take a more pro-active government stance for organic farming (*or genetic engineering*)". Table 2 provides two comparisons of professionals' views on the need for a more pro-active government policy stance: the general sample and the total sample.

As hypothesised, the more positive a respondent was towards an innovation, the more they advocated a stronger governmental stance. For example, proponents of organic farming and genetic engineering were significantly more likely to agree with the question (mean 4.03

and 3.41 respectively) than opponents (mean 2.59 and 1.91 respectively) (t tests: $p<0.000$ for both).[7]

The second hypothesis (H19) to be examined is the idea that professionals who are favorable to an alternative paradigm are more likely to name 'inside scientific' sources as innovation barriers than 'outside public' sources than professionals negative to an alternative paradigm. Professionals were asked: "What do you think are the major barriers facing the diffusion of genetic engineering (*or organic farming*) in Australia?" Responses were categorised as (1) inside scientific barriers (i.e. lack of scientific knowledge, uncertainty about future benefits/costs, lack of independent verification, scientists' attitudes/perceptions; misinformation by scientists, no long-term scientific studies conducted, information needs, lack of scientific research and development, and lack of extension/scientific support); (2) outside public barriers (i.e., media portrayal, public attitudes/perception/knowledge, lobby group influence, and political regulations); (3) practical innovation issues (market, on-farm issues); and (4) other. The results for (1) and (2) for the total sample are shown in Table 3.

Professionals who were favorable to organic farming were much more likely to name inside sources as barriers to further diffusion than those that felt negatively (27.4% vs 10.7% of the total barriers named, respectively) but were marginally more likely to name outside public barriers than negative respondents (3.3% vs 1.0% respectively). However, those that felt favourably towards genetic engineering were much less likely to name inside barriers than those that felt negatively (11.9% vs 35.3% respectively), and were more likely to name outside barriers than negative respondents (61.5% vs 32.4% respectively).

Table 3. Percentage of Inside & Outside Barriers out of Total Barriers Named by Innovation for the Total Sample (n = 185) %

Type of Barrier Named		Positive	Unsure	Negative	Total
Inside Scientific	Organic Farming	27.4	24.1	10.7	22.7
	Genetic Engineering	11.9	22.0	35.3	23.1
Outside Public	Organic Farming	3.3	2.3	1.0	2.6
	Genetic Engineering	61.5	39.4	32.4	44.4

Table 4. Changes in Agricultural Professionals Views by Innovation

Questions on Agricultural Innovations	Genetic Engineering Answers			Organic Farming Answers		
	General	Target	Total	General	Target	Total
Views becoming g more favourable over the past 5 yrs Views becoming less favourable over the past 5 yrs Plans to learn more about the innovation in the future	25% 16% 83%	20% 32% 80%	23% 22% 82%	35% 13% 76%	46% 6% 89%	38% 10% 81%

[7] All t tests were conducted with a student two-tailed distribution.

Table 5. Summary of Hypothesis Tests for a Paradigm Shift in Agriculture

Paradigm Hypothesis	Hypothesized Response	Organic Farming Actual Sign	Organic Farming Result	Genetic Engineering Actual Sign	Genetic Engineering Result
Status					
H1: Younger professionals	+	(-)[1]		-	✗
H2: Professionals with more working experience in area	-	-	✓	(-)	
H3: High- status professionals.	+	(+)		(-)	
Self-Interest					
H4: Professionals who have a higher salary	-	(-)		(-)	
H5: Professionals with higher alternative research experience	+	(+)		+	✓
H6: Professionals who have increased years of tertiary educat.	-	+	✗	(-)	
H7: Professionals with higher alternative paradigm knowledge	+	+	✓	(-)	
Specialisation					
H8: Professionals who are scientists	-	(+)		+	✗
H9: Professionals who are extension officers	-	(-)		+	✗
H10: The existence of any occupational effects for organisation	-	-	✓	(+)	
H11: Professionals whose job is less specialised	+	(+)		(-)	
Gender					
H12: Female professionals	+	(+)		(-)	
Information Choices					
H18: Professionals gain info. primarily from scientific sources	-	-	✓	(-)	
H19: Professionals name inside more than outside sources	+	+	✓	-	✗
Changes in Views					
H20: Conversion of professionals to the new paradigm	+	+	✓	+	✓
Perceptions					
H13: Professionals views on profitability	+	+	✓	+	✓
H14: Professional views on the sustainability of current agric.	+	(+)		-	✗
Principles					
H15: Professionals who advocate the need for an increased government policy	+	+	✓	+	✓
H16: Professional views on intellectual property rights in agricultural research	-	(+)		+	✗
H17: Professionals views on private funding issues in research	-	(+)		-	✓
Simple Count	✓	8	✓	5	
	✗	1	✗	6	

Note: 1. Signs in brackets indicate the result was not statistically significant at 10% (or below) level.

Overall, professionals were more likely to name inside scientific sources (22.7%) as organic barriers than outside public barriers (2.6%). Conversely, professionals were much more likely to name outside public sources (44.4%) than inside scientific sources (23.1%) as genetic engineering barriers. The quantitative evidence for H19 is therefore much stronger for organic farming.

The final individual hypothesis test (H20) is to consider how professional views on organic farming and genetic engineering are changing. For a paradigm shift in agriculture to occur, then a prerequisite must mean that agricultural professionals have to be "converted" to the new paradigm (Kuhn 1970). Such a hypothesis is extremely hard to test without comparable historical data for Australia. To assess how these professional attitudes have changed recently, two questions were asked: "Over the last 5 years, have your views on organic farming *(or genetic engineering)* become (a) more favorable; (b) less favorable; or (c) stayed the same?" Table 4 shows the results for both samples, which is derived from Wheeler (2007). Professionals were statistically more likely to have positively changed their opinion of organic farming and genetic engineering than negatively (35% vs 13% and 25% vs 16% respectively). Both innovations therefore provide statistical support for H20, though overall there has been more net favorable change in organic than genetic engineering views.

Discussion

Table 5 displays a summary of the hypotheses tests. A tick in Table 5 indicates the hypothesis had the right sign for a new paradigm test and was statistically significant. A cross indicates the hypothesis had the wrong sign and was statistically significant, while a blank indicates either the incorrect sign or statistical insignificance.

Now, the question remains as to how much evidence is required before there is support for an agricultural paradigm shift. Given that a paradigm shift occurs very rarely in science, it is to be expected that applying these tests to potential candidates would mean most would fail. How can we distinguish between an innovation and a paradigm? In other words, what results could we expect from a new theory or method in agriculture that is a new innovation but not a new paradigm? We argue that if a new theory is an innovation only, then applying these tests should result mainly in a lack of statistical significance. There should be no clear pattern among the way that professionals react to new ideas. If there is statistical significance, then it is to be expected that such significance should be randomly distributed and hence there should be an equal chance of satisfying the individual hypothesis. A systematic pattern of reactions should only emerge if professionals are consistently reacting the way the literature suggests. Having said this, it is another thing entirely to suggest how many hypotheses need to be satisfied. Probably at least half the hypotheses need to be met, but the question remains open as to if this is enough. For a new theory to be a new paradigm, it may have to meet most of the hypotheses, while weaker evidence may suggest a sub-paradigm, or even less evidence may suggest a new research program.

Table 5 shows there is statistical support for eight of the organic farming hypotheses, with six others having the expected sign for a paradigm test but were not statistically significant. Only one hypothesis (education effect) in the organic model had the wrong sign for a paradigm test but was statistically significant (see discussion later). These results may indicate evidence to suggest that organic farming fits the mould of a potential new paradigm in agriculture.

There is far less evidence that genetic engineering represents a new paradigm in agriculture, with statistical support for only five hypotheses, while four had the expected sign but were statistically insignificant. Six hypotheses had the wrong sign for a paradigm test and were statistically significant. These results seem to suggest a certain element of randomness in the statistical significance of the genetic engineering hypotheses.

The case for organic farming becomes stronger when we consider the education hypothesis. The hypothesis suggested that increased education would have a negative effect on views towards a new paradigm. However, education could indeed have the reverse effect on views if the Milgram effect dominated and professionals viewed education as a sunk cost. Education is found to be a positive influence in the organic farming model and a negative influence in the genetic engineering model. Given the other results from the models, it is likely that the Milgram effect does indeed dominate, which means that the expected sign on the education paradigm hypothesis is most likely positive, not negative. Increased tertiary education may play a role in encouraging professionals to critically assess current scientific thinking, as well as encouraging a skeptical mind. If this supposition is correct, then there is even stronger statistical evidence for organic farming to represent a new paradigm.

The age result is also interesting. Age was a positive influence on innovation acceptance in both models (though only significant for genetic engineering). These results do not support the generalised hypothesis that younger scientists are more likely to accept radical new ideas, while they support the idea that other factors (that may be correlated with age) play a more important role. Working age seems to be the more important variable, especially for organic farming.

Implications of the Presence of a Dominant Paradigm in Agriculture

There seems to be some statistical support that organic farming may represent a new agricultural paradigm, with far less support for genetic engineering. What can we learn from such a result? Does this indicate a strong conclusion that organic farming may become a dominant paradigm in agriculture in the future? A weaker conclusion may be that there are a number of sub-paradigms (or a number of research programmes) in agriculture, and organic farming is on the way to establishing itself as one such alternative research programme. It can be concluded that agricultural genetic engineering is already well-established as a research programme in agriculture, one that is different and innovative but still shares many common fundamental or 'hardcore' assumptions as the current dominant paradigm (conventional agriculture). Organic farming as a research programme also shares many similar skills and assumptions, but differs on fundamental assumptions. It is not completely mutually exclusive to the current paradigm of conventional agriculture but it is different. Therefore, organic farming has more characteristics to qualify as a sub-paradigm in agriculture than genetic engineering does.

The presence of paradigms may result in a number of problems (such as indirect scientific suppression, the delay of new ideas and the emergence of path dependence). Many writers have highlighted examples of agricultural path dependence (Wilson & Tisdell 2001, Wolffe & Recke 2000, Ison & Russell 2000, Cowan & Gunby 1996, MacIntyre 1987). It may also have serious implications for the adoption of organic farming. If organic farming indeed represents a potential new paradigm in agriculture, then this may have implications for agricultural research and infrastructure. Such research may need to change from its current conventional focus to a system of research that is more holistic, long-term, multi-disciplinary, externality encompassing, on-farm and region specific (such as described in MacRae *et al.*

1989 and 1990). At the very least, there is support for the argument that organic farming should receive more research allocation. Also, government policies, education and extension may need to change.

Conclusion

The current conflict in agriculture exists at both levels of paradigmatic debate, societal and scientific. Organic farming and genetic engineering represent two innovations in agriculture that had some *prime a facie* suggestion of being new paradigms. This paper uses Kuhn's analogy of a paradigm shift in science for a paradigm shift in agriculture, though they are not strictly the same thing. Qualitative evidence from a range of different writers provides support for the concept of a paradigm shift in agriculture towards organic farming, but less support for genetic engineering which has more characteristics of an innovation than a new paradigm.

This paper put forward a number of hypotheses to test whether agricultural professionals' reactions to organic farming and genetic engineering fit a predetermined response to a new paradigm. Such hypotheses are not considered to be conclusive in any way, and further work will be needed to establish tests for detecting paradigms. Using empirical results from previous studies, this paper has shown there is relatively strong quantitative evidence to support organic farming as a new paradigm, but only very weak quantitative evidence for genetic engineering. Further bivariate ordered probit regression modeling confirmed that professionals are reacting differently to the two innovations, which provides further evidence of the stability of the results discussed here. The evidence may actually support the presence of research programmes in agriculture rather than them being paradigms (in the strict sense). Therefore, the detection of organic farming as a new research programme or a new sub-paradigm does not necessarily mean organic farming will ever become the 'dominant paradigm' in agriculture, it may just remain a relatively small research programme that is an alternative to more dominant sub-paradigms. At the very least, the results in this study support a change in agricultural research, funding, education and dissemination. Research may need to move from its current conventional focus to a system that is more holistic, long-term, multi-disciplinary, externality encompassing, on-farm and region specific.

References

Abaidoo, S. & Dickinson, H. (2002). Alternative and conventional agricultural paradigms: Evidence from farming in southwest Saskatchewan, *Rural Sociology*, Vol. *67*, 114-131.

Allen, T. (1969). The Differential Performance of Information Channels in the Transfer of Technology in W. Gruber, & D. Marquis, (Ed) *Factors in the Transfer of Technology*, MIT Press, Massachusetts.

Allen, P. (1996). "Social imperatives in organic agriculture: a reflection", in ∅stergaard, T. (Ed) (1996) *Fundamentals of Organic Agriculture*, 11th IFOAM International Scientific Conference August 11-15, 1996, Copenhagen.

Avery, D. (1995). *Saving the Planet with Pesticides and Plastic: The Environmental Triumph of High-Yield Farming*, Hudson Institute, Indianapolis, Indiana.

Barnes, B., Bloor, D. & Henry, J. (1996). *Scientific Knowledge: a Sociological Analysis*, University of Chicago Press, Chicago.

Barnes, B. & MacKenzie, D. (1979). On the Role of Interests in Scientific Change, in R. Wallis, (Ed) *On the Margins of Science: The Social Construction of Rejected Knowledge*, University of Keele, Keele.

Becker, H. & Carper, J. (1956). The elements of identification with an occupation, *American Sociological Review*, Vol. *21*, 341-348.

Beus, C. & Dunlap, R. (1990). Conventional versus Alternative Agriculture: The Paradigmatic Roots of the Debate, *Rural Sociology*, Vol. *55*, 590-616.

Beyer, P., Al-Babili, S., Ye, X., Lucca, P., Schaub, P., Welsch, R. & Potrykus, I. (2002). Golden Rice: Introducing the Beta-Carotene Biosynthesis Pathway into Rice Endosperm by Genetic Engineering to Defeat Vitamin A Deficiency, *Journal of Nutrition*, Vol. *132*, 506-510.

Blaug, M. (1975). Kuhn versus Lakatos, or paradigms versus research programmes in the history of economics, *History of Political Economy*, Vol. *7*, 399-433.

Borlaug, N. (2000). Ending world hunger. The promise of biotechnology and the threat of antiscience zealotry, *Plant Physiology*, Vol. *124*, 487-490.

Broad, W. & Wade, N. (1982) .*Betrayers of the Truth: Fraud and deceit in the Halls of Science*, Simon and Schuster, New York.

Burr, C. (2002). *The Emperor of Scent: A Story of Perfume, Obsession, and the Last Mystery of the Senses*, Random House, New York.

Busch, L. & Lacy, W. (1983). *Science, Agriculture and the Politics of Research*, Westview Press Inc, Boulder.

Buttel, F. & Goldberger, J. (2002). Gender and Agricultural Science: Evidence from Two Surveys of Land-Grant Scientists, *Rural Sociology*, Vol. *67*, 24-45.

Cowan, R. & Gunby, P. (1996). Sprayed to Death: Path dependence, Lock-in and Pest Control Strategies, *The Economic Journal*, Vol. *106*, 521-542.

K. Dahlberg, (Ed) (1986). *New Directions for Agriculture and Agricultural Research: Neglected Dimensions and Emerging Alternatives*, Rowman & Allanheld Publishers, New Jersey.

David, P. (2000). The Political economy of public science, in H. Smith, (ed) *The Regulation of Science and Technology*, Macmillan, London.

A. de Grazia, (Ed) (1966). *The Velikovsky Affair*, Sidgwick & Jackson, London.

Dimara, E., Petrou, A. & Skuras, D. (2003). The socio-economics of niche market creation: A social ecology paradigm for the adoption of organic cultivation in Greece, *International Journal of Social Economics*, Vol. *30*, 219-235.

Dolby, R. (1979). "Reflections on Deviant Science", in R. Wallis, (Ed) *On the Margins of Science: The Social Construction of Rejected Knowledge*, University of Keele, Keele.

Drahos, P. (1996). *A Philosophy of Intellectual Property*, Dartmouth Publishing Company Ltd, Aldershot.

Egri, C. (1994). *Power, Politics and the Innovation Process: Analysis of an Organisational Field in Agriculture*, PhD thesis, Department of Commerce, University of British Columbia.

Ewens, S. & Pusztai, A. (1999). Effect of Diets Containing Genetically Modified Potatoes Expressing Galanthus Nivalis Lectin on Rat Small Intestine, *Lancet*, Vol. *354*, 1353-1354.

Feyerabend, F. (1988). *Against Method*, Verso, London.

Gepts, P. (2002). A Comparison between Crop Domestication, Classical Plant Breeding, and Genetic Engineering, *Crop Science*, Vol. *42*, 1780-1790.

Hadwiger, D. (1982). *The Politics of Agricultural Research*, University of Nebraska Press, London.

Harp, A. & Sachs, C. (1992). Public agricultural researchers: Reactions to organic, low input and sustainable agriculture, *Agriculture and Human Values*, Vol. *9*, 58-63.

Hopper, K. & Thorburn, L. (2005). *2005 Bio-Industry Review: Australia & New Zealand*, Innovation Dynamics.

Huffman, W. (2001). Public-private research and development relationships: discussion, *American Journal of Agricultural Economics*, Vol. *83*, 754-757.

Huffman, W. & Tegene, A. (2002). Public Acceptance of and Benefits from Agricultural Biotechnology: A Key Role for Verifiable Information, in V., Santaniello R. Evenson, & D. Zilberman, (Eds). *Market Development for Genetically Modified Foods*, CABI Publishing, Wallingford.

Independent Science Panel, (2003). *The Case for a GM-Free Sustainable World*, published by Institute of Science in Society, available at http://www.indsp.org/ISPreportSummary.php

Ison, R. & Russell, D. (Ed) (2000). *Agricultural Extension and Rural Development: Breaking out of Traditions*, Cambridge University Press, Cambridge.

ISSC, (1993). Food Biotechnology: A New Paradigm for Food, the Farm, and the Public, *Bio/Technology*, Vol. *11*, 1584-1589.

King, D. (2004). *GM Science Review: Second Report*, GM Science Review Panel for the UK Govt, January.

Krimsky S. (2000). *Hormonal Chaos: The Scientific and Social Origins of the Environmental Endocrine Hypothesis*, The Johns Hopkins University Press, Baltimore.

Kuhn, T. (1970). *The Structure of Scientific Revolutions*, The University of Chicago Press, Chicago.

Latour, B. & Woolgar, S. (1986). *Laboratory Life*, Princeton University Press, New Jersey.

Latour, B. (1987). *Science in Action: How to Follow Scientists and Engineers through Society*, Open University Press, Milton Keynes.

Liebowitz, S. & Margolis, S. (1990). "The Fable of the Keys", *Journal of Law and Economics*, Vol. XXXIII, 1-25.

Liebowitz, S. J. & Margolis, S. E. (1994). 'Network Externality: An Uncommon Tragedy', *Journal of Economic Perspectives*, Vol. *8*, 133-150.

Macilwain, C. (2004). "Organic: Is it the future of farming?" *Nature*, Vol. *428*, 792-793.

MacIntyre, A. (1987). "Why Pesticides Received Extensive Use in America: A Political Economy of Agricultural Pest Management to 1970", *Natural Resources Journal*, Vol. *27*, 533-578.

MacRae, R, Hill, S, Henning, J. & Mehuys, G. (1989). Agricultural Science and Sustainable Agriculture: a Review of the Existing Scientific Barriers to Sustainable Food Production and Potential Solutions, *Biological Agriculture and Horticulture*, *6*, 173-219.

MacRae, R, Hill, S, Henning, J. & Bentley, A. (1990). Policies, programs, and regulations to support the transition to sustainable agriculture in Canada, *American Journal of Alternative Agriculture*, *5*, 76-92.

Mäder, P., Fliebach, A., Dubois, D., Gunst, L., Fried, P. & Niggli, U. (2002). "Soil fertility and biodiversity in organic farming", *Science*, Vol. *296*, 1694-1697.

Manwell, C. & Baker, C. (1988). "Information Flow in Agriculture: The Major Interest Groups and their Interactions", *Philosophy and Social Action*, Vol. *14*, 51-64.

Martin, B. (1979). *The Bias of Science*, Southwood Press Pty Ltd, Marrickville, Australia.

Mazzoleni, R. & Nelson, R. (1998). "The Benefits and Costs of Strong Patent Protection: A Contribution to the Current Debate", *Research Policy*, Vol. *27*, 273-284.

Merrill, M. (1983). "Eco-Agriculture: A Review of its History and Philosophy", *Biological Agriculture and Horticulture*, Vol. *1*, 181-210.

Merton , R. (1973). *The Sociology of Science: Theoretical and Empirical Investigations*, The University of Chicago Press, Chicago and London.

Milgram, S. (1983). *Obedience to Authority: An Experimental View*. New York: Harper/Collins.

Miller, H. (1993). "Perception of Biotechnology Risks: The Emotional Dimension", *Bio/Technology*, Vol. *11*, 1075-1076.

Mulkay, M. (1972). *The Social Process of Innovation: A Study in the Sociology of Science*, Macmillian, London.

Mulkay, M. (1979). *Science and the Sociology of Knowledge*, Allen & Unwin, London.

National Academy of Sciences, (2000). *Transgenic Plants and World Agriculture,* National Academies Press, Washington.

O'Riordan, T. & Cobb, D. (2001). "Assessing the Consequences of Converting to Organic Agriculture", *Journal of Agricultural Economics*, Vol. *52*, 22-35.

Popper, K. (1970). "Normal Science and its Dangers", in I. Lakatos, & A. Musgrave, (eds) (1970) *Criticism and the Growth of Knowledge*, Cambridge University Press, London.

Prakash, C. (2001). "The genetically modified crop debate in the context of agricultural evolution", *Plant Physiology*, Vol. *126*, 8-15.

Pretty, J. (1995). "Participatory Learning for Sustainable Agriculture", *World Development*, Vol. *23*, 1247-1263.

Price, D. (1969). "The Structures of Publication in Science and Technology" in W. H. Gruber, D. G. Marquis, (Ed) *Factors in th e Transfer of Technology*, The MIT Press, Massachusetts.

Quist,, D. & Chapela, I. (2001). "Transgenic DNA introgressed into traditional maize landraces in Oaxaca, Mexico", *Nature*, Vol. *41*, 541-543.

Riggs, P. (1992). *Whys and Ways of Science: Introducing Philosophical and Sociological Theories of Science*, Melbourne University Press, Melbourne.

Shepherd, M., Pearce B., Cormack B., Philipps L., Cuttle S., Bhogal A., Costigan P. & Unwin R. (2003). *An Assessment of the Environmental Impacts of Organic Farming*. DEFRA-funded project OF0405, DEFRA, London.

Stolze, M., Piorr, A., Haring, A. & Dabbert, S. (2000). "The environmental impacts of organic farming in Europe", Organic Farming in Europe, Economics and Policy, 6, University of Hohenheim: Stuttgart-Hohenheim.

Tisdell, C. (1991). *Economics of Environmental Conservation: Economics for Environmental and Ecological Management*, Elsevier, Amsterdam.

Trewavas A. (2001). "Urban myths of organic farming", *Nature*, Vol. *410*, 409-410.

Turner, L. (2004). "Biotechnology as religion", *Nature Biotechnology*, Vol. *22*, 659-660.

van den Bosch, R. (1978). *The Pesticide Conspiracy: An alarming look at pest control and the people who keep us "hooked" on deadly chemicals*, Doubleday & Company Inc, Garden City, New York.

Wheeler, S. (2007). "Contrasting the beliefs of Australian agricultural professionals about the benefits and costs of genetic engineering and organic agriculture", *Australian Journal of Experimental Agriculture*, Vol. *47*, 1389-1396.

Wheeler, S. (2008). "What Influences Agricultural Professionals' Views towards Organic Agriculture?" *Ecological Economics*, Vol. *65*, 145-154.

Wheeler, S. (pending). "Exploring the Influences on Australian Agricultural Professionals' Genetic Engineering Beliefs: An Empirical Analysis", *Journal of Technology Transfer*, (DOI: 10.1007/s10961-008-9094-y).

Wilson, S. & Barnes, I. (1995). "Scientists' Participation in Environmental Policy", *Search*, Vol. *26*, 270-273.

Wilson, C. & Tisdell, C. (2001) "Why farmers continue to use pesticides despite environmental, health and sustainability costs", *Ecological Economics*, Vol. *39*, 449-462.

Wolffe, H. & Recke, G. (2000). "Path Dependence and Implementation Strategies for Integrated Pest Management", *Quarterly Journal of International Agriculture*, Vol. *39*, 149-171.

Wynen, E. (1996). "Research Implications of a Paradigm Shift in Agriculture: the case of Organic Farming" *Resource and Environmental Studies*, 12, Centre for Resource and Environmental Studies, ANU.

Wynen, E. (2003). *Organic Agriculture in Australia – Levies and Expenditure*, report prepared for RIRDC, No. 03/2002, March

APPENDIX ONE

Explanatory Variable Table

Variable	Explanation
Age Working Age Ethnicity Gender Farm Background Tertiary Education Years Salary	Actual age in years squared Years spent working in current agricultural area Dummy variable, 1 = non-European, 0 = European Dummy variable, 1 = male, 0 = female Dummy variable, 1 = farm background, 0 = otherwise Years of post-secondary education Salary of professionals (midpoint of salary band of job classification level)
Scientist Leader Extension role Natural resource focus CSIRO	Dummy variable, 1 = scientist, 0 = otherwise Dummy variable, 1 = leader, 0 = otherwise Dummy variable, 1 = extension role, 0 = otherwise Dummy variable, 1 = natural resource management focus, 0 = otherwise Dummy variable, 1 = employed at CSIRO, 0 = otherwise
Target OA/GE knowledge OA/GE research relevance Media main source info for GE Scientific info main source for OA/GE OA farms main source info	Dummy variable, 1 = targeted sample, 0 = otherwise Scalar variable of OA (or GE) knowledge, where 1= no knowledge and 5 = detailed and practical knowledge % indicating the research relevance of OA (or GE) in the past 3 years Dummy variable, 1 = media main source for GE info, 0 = otherwise Dummy variable, 1 = scientific information (journals, peers, conferences etc) as first main source for OA (or GE) information, 0 = otherwise Dummy variable, 1 = organic farms first main source for info, 0 = otherwise

Views on CA's sustainability	Scalar variable of response to "Conv. agric. farming in Aus. is generally not environment. sustainable", where 1= strongly disagree to 5 = strongly agree
OA/GE benefit-cost ratio	Ordered dependent variable of response to "do you think the benefits of OA/ GE are greater than their costs/risks?", where 0 = no, 1 = unsure, 2 = yes
Views on OA profitability Views on OA environmental sustainability Views on OA product quality Views on OA innovativeness Views on OA yields	Scalar variable of response to the statement that "CA is more financially profitable than OA", where 1= strongly disagree and 5 = strongly agree Scalar variable of response to the statement that "OA is more environmentally friendly than CA", where 1= strongly disagree and 5 = strongly agree Scalar variable of response to the statement that "OA produces better quality of food than CA", where 1= strongly disagree and 5 = strongly agree Scalar variable of response to the statement that "OA farming is a return to pre-1950s agriculture", where 1= strongly disagree and 5 = strongly agree Dummy variable where 1 = belief that conversion from CA to OA results in falls of yield over 30
Views on agric. IPR and Patents Views on agric. private funding effects	Scalar variable of response to "the increased use of intellectual property rights & patenting activity is a good thing for agric. research", of 1 to 5 Scalar variable of response to the statement that "private funding of agricultural research does not influence research outcomes", of 1 to 5
Views on GE profitability Views on GE environmental sustainability Views on GE product quality Views on GE yields Views on GE long-term research needs	Scalar variable of response to the statement that "GE is more financially profitable than CA", where 1= strongly disagree and 5 = strongly agree Scalar variable of response to the statement that "GE is more environmentally friendly than CA", where 1= strongly disagree and 5 = strongly agree Scalar variable of response to the statement that "GE produces better quality of food than CA", where 1= strongly disagree and 5 = strongly agree Dummy variable where 1 = belief that conversion from CA to GE results in increases of yields over 30% Scalar variable of response to "there needs to be more research on LT consequences of GE products before further general release", of 1 to 5

Bivariate Ordered Probit Regression Results

	Coef.	*z*	*P>|z|*
GENETIC ENGINEERING			
Age	**-0.14**	**-1.84**	**0.07**
Age^2	**0.00**	**1.97**	**0.05**
Working Age	**-0.01**	**-0.55**	**0.59**
Ethnicity	**1.52**	**2.13**	**0.03**
Female	**0.22**	**0.98**	**0.33**
Having a Farm Background	**-0.51**	**-2.66**	**0.01**
Years of Tertiary Education	**-0.05**	**-0.74**	**0.46**
Salary	**0.00**	**0.75**	**0.45**
Job title is Scientist	**0.62**	**2.55**	**0.01**
Classified as a Leader in their Position	**-0.05**	**-0.15**	**0.88**
Working predominantly as an Extension Officer	**0.39**	**1.67**	**0.09**
Employed by a state organization	**0.15**	**0.47**	**0.64**
Part of the Targeted Sample	**-0.13**	**-0.55**	**0.58**
Working in the area of Natural Resource Management	**-0.08**	**-0.31**	**0.76**

Bivariate Ordered Probit Regression Results (continued)

	Coef.	*z*	*P>|z|*
GENETIC ENGINEERING			
Working in the area of Broadacre and/or Grazing	**0.13**	**0.50**	**0.62**
Employed by CSIRO	**0.14**	**0.28**	**0.78**
Genetic Engineering Knowledge	**-0.06**	**-0.46**	**0.64**
% of Research Time Relevant to Genetic Engineering	**0.01**	**2.10**	**0.04**
Views on Conventional Agriculture's Sustainability	**-0.24**	**-2.62**	**0.01**
Views on agric. intellectual property rights & patents	**0.23**	**2.37**	**0.02**
Views on private funding of agricultural research	**-0.15**	**-1.51**	**0.13**
Views on Genetic Engineering's Profitability	**0.39**	**2.69**	**0.01**
Beliefs on uncertainty of GE and need for long-term research	**-0.65**	**-4.59**	**0.00**
Views on Genetic Engineering Food Quality	**0.11**	**0.80**	**0.43**
Views on Genetic Engineering Environmental Sustainability	**0.50**	**3.48**	**0.00**
Naming the Media as the first main source of Info.	**0.28**	**1.11**	**0.27**
Naming Scientific sources as first main source of Info.	**-0.08**	**-0.28**	**0.78**
Age	**-0.13**	**-1.50**	**0.13**
Age^2	**0.00**	**1.63**	**0.10**
Working Age	**-0.03**	**-1.68**	**0.09**
Ethnicity	**0.60**	**1.14**	**0.25**
Female	**-0.18**	**-0.65**	**0.52**
Having a Farm Background	**-0.03**	**-0.13**	**0.90**
Years of Tertiary Education	**0.13**	**2.10**	**0.04**
Salary	**0.00**	**-0.11**	**0.91**
Job title is Scientist	**0.00**	**0.01**	**0.99**
Classified as a Leader in their Position	**0.19**	**0.48**	**0.63**
Working predominantly as an Extension Officer	**-0.06**	**-0.24**	**0.81**
Employed by a state organization	**0.02**	**0.06**	**0.95**
Part of the Targeted Sample	**0.64**	**1.90**	**0.06**
Working in the area of Natural Resource Management	**0.35**	**1.13**	**0.26**
Working in the area of Broadacre and/or Grazing	**0.04**	**0.13**	**0.89**
Employed by CSIRO	**-0.88**	**-1.77**	**0.08**
Organic Agriculture Knowledge	**0.24**	**1.76**	**0.08**

% of Research Time Relevant to Organic Farming	**0.01**	**1.14**	**0.25**
Views on Conventional Agriculture's Sustainability	**0.12**	**1.10**	**0.27**
Views on agric. intellectual property rights & patents	**0.09**	**0.82**	**0.41**
Views on private funding of agricultural research	**0.02**	**0.15**	**0.88**
Views on Organic Farming's Profitability	**0.69**	**5.07**	**0.00**
Views on Organic Farming's Innovativeness	**0.28**	**2.36**	**0.02**
Views on Organic Farming's Food Quality	**-0.36**	**-2.68**	**0.01**
Views on Organic Farming's Environmental Sustainability	**-0.25**	**-1.57**	**0.12**
Naming the Media as the first main source of Info.	**-1.15**	**-3.55**	**0.00**
Naming Scientific sources as first main source of Info.	**-0.35**	**-1.32**	**0.19**
/cut11	**-3.38**	**1.99**	
/cut12	**-2.05**	**2.00**	
/cut21	**-0.55**	**1.94**	
/cut22	**0.75**	**1.95**	
Number obs.	**185.00**		
Log Likelihood	**-252.59**		
Wald Chi2	**128.85**		
Prob>chi2	**0.00**		
rho	**0.08**	**0.14**	
Wald test of dep. Equns chi2(1)	**0.30**		
Prob > chi2	**0.58**		

In: Organic Farming and Peanut Crops
Editors: D. C. Grossman, T. L. Barrios, pp. 191-202
ISBN: 978-1-60876-187-6

Chapter 8

ORGANIC PRODUCE, PRICE PREMIUMS, AND ECO-LABELING IN U.S. FARMERS' MARKETS

***Amy Kremen*[1]*, Catherine Greene*[1] *and Jim Hanson*[2]**
[1]College of Agriculture and Natural Resources, The University of Maryland, USA
[2]Resource Economics Division, Economic Research Service

ABSTRACT

The number and popularity of farmers' markets in the United States have grown concurrently with organic production and consumer interest in locally and organically produced foods. This research describes the significance of farmers' markets as market outlets for many organic farmers, and recent shifts in relationships between organic growers, market managers, and customers. Market managers in more than 20 States answered questions by phone pertaining to the 2002 market season. Their responses provide insight into recent grower, manager, and customer decision making and attitudes about foods advertised and sold as organic at farmers' markets. Specific areas of interest include participation of organic farmers in markets, demand for organic products, advertising strategies used to highlight organic products, and price premiums charged for organic products.

INTRODUCTION

Farmers' markets have been growing in number and popularity over the past decade, providing valuable opportunities for thousands of full- and part-time farmers. While the volume of produce sales via farmers' markets is small—less than 2 percent of U.S. sales overall (Handy et al.)—the number of farmers' markets in the U.S. has increased by 79 percent since 1994 to more than 3,100 in 2002. Consumers representing a wide range of socioeconomic backgrounds are demonstrating growing interest in and support for local agriculture (Stephenson and Lev).

Participation by growers at farmers' markets depends on several factors, including a market's popularity, location, and management. Customer participation depends primarily on a market's location, since most customers tend to shop at markets close to where they live (Brown). Ultimately, markets that attract and retain vendors and shoppers are those that satisfy the needs of both (Brown). Market success can depend on how attuned managers are to customer demand and farmers' interests, and on how well they can satisfy these needs.

Previous studies indicate that freshness, high quality, fair pricing, pleasant social interaction with farmers and market shoppers, and locally grown foods are attributes commonly sought by customers attending farmers' markets (Lockeretz, Brown). Vendors at farmers' markets also frequently provide heirloom and specialty produce varieties, hand-crafted products, and other items valued by customers but difficult to mass-produce and are rare at other markets (Lyson et al.). This study shows that many customers at farmers' markets also appreciate having direct access to farmers that use ecologically sensitive agricultural techniques—such as organic production methods—on their farms.

Research on the marketing habits of organic growers indicates that they tend to use direct marketing outlets, such as farmers' markets, much more often than the general U.S farmer population (Walz, Fernandez et al.). Although the marketing channels for organic food have expanded in recent years, farmers' markets remain an accessible, low-cost, and flexible sales outlet for many organic producers. Insufficient infrastructure—a lack of established purchasing, storage, and distribution channels, for example—may still hinder growers interested in expanding or coordinating their production to serve other local market outlets (Slama, Birkerts). Several market managers interviewed for this study indicated that organic farmers routinely play a significant role in creating farmers' markets, and often serve as manager or board member.

For this study, farmers' market managers in more than 20 States answered questions by phone pertaining to the 2002 market season. Market managers were interviewed in late 2002 and early 2003 about participation by organic growers and customer demand for organic products. Questions about the 2002 market season (14 total) were modeled on the market manager questionnaire being used in an ongoing study of U.S. farmers' markets by USDA's Agricultural Marketing Service. The AMS study, initially conducted in 1994 and repeated in 2000, surveys all farmers' market managers in the United States, and has shown that farmers' markets are "a growing marketing tool for farmers" and that "customers can benefit from direct contact with the producers of their food" (Payne).

The AMS study collects a wide range of information on farmer and customer demographics and market characteristics. This study focuses more narrowly on the role of farmers' markets as a marketing tool for organic growers and on current customer attitudes about organic products at these markets.

The group of market managers interviewed was selected with guidance from AMS-appointed State farmers' market representatives who maintain lists of managers and markets. State representatives were asked to provide a list of approximately 15 markets that would represent the diversity in size, "feel," and location of their State's markets. Interviews were conducted with 210 market managers located in 23 States and Washington, DC (map 1). Although market sampling was not random, the data generated provide useful information for understanding the current role of organic growers at markets within a wide range of communities. The data collected in this study for many market characteristics—including market type, average levels of farmer attendance, and customer distances traveled to market—correspond closely to data collected for those characteristics by AMS in its much larger study of U.S. farmers' markets.

Organic Farmers Participate Widely in U.S. Farmers' Markets

Organic growers participated at more than four-fifths of the 210 markets in this study, and represented approximately one-third of regularly attending farmers during the 2002 season (table 1). Such broad participation highlights the disproportionately high use of farmers' markets as a sales outlet by organic growers. While only one or two organic growers participated in about half of the markets with organic growers, 11 percent of these markets had more than ten participating organic growers (table 1). Many of these farmers had already acquired third-party organic certification, or had developed alternative eco-labels to promote their products, even though Federal rules requiring organic certification had not yet taken effect (see box).

Table 1. Organic grower participation in 210 U.S. farmers' markets,2002

Organic farmer participation	
Number of farmers in 210 farmers'markets	4,681
Number of participating organic farmers	1,416
Markets with organic farmers (percent)	82
Number of organic farmers present (percent):	
1-2 growers	46
3-10 growers	43
11-40 growers	11

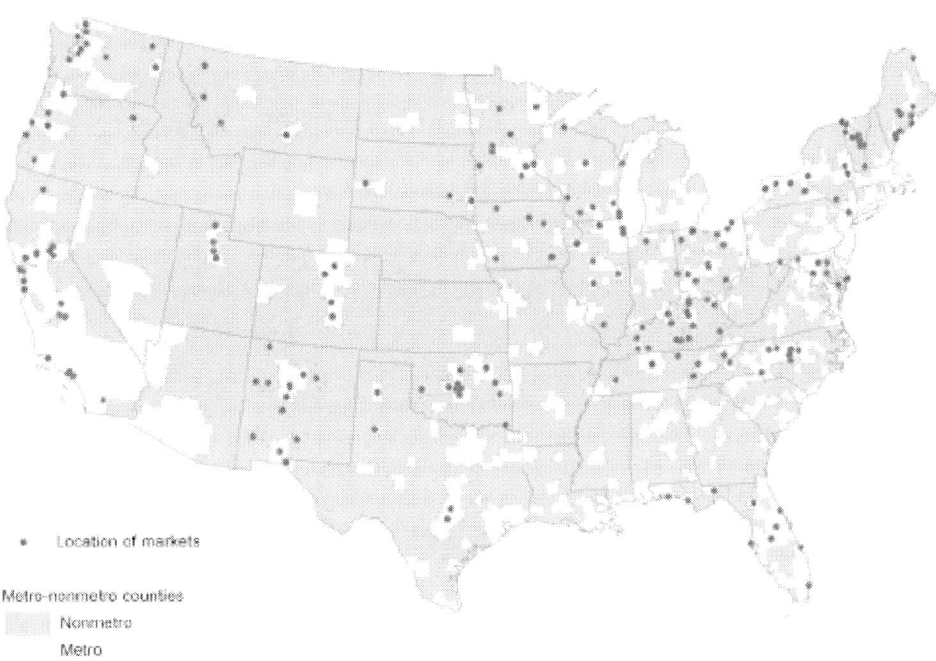

Map 1. Distribution of 210 U.S. farmers' markets interviewed, 2002

The data collected in this study are not conclusive about whether organic growers are more likely to attend urban (metro) or rural (nonmetro) farmers' markets. Organic farmer participation was sometimes limited or negligible at several large, highly popular urban markets, but was significant at several smaller metro and rural markets, a few of which had exclusively organic vendors.

Managers were able to provide general estimates of gross sales (not necessarily limited to sales at the market) of the organic farmers participating at the markets in this study. Nearly 90 percent of the certified organic growers, and slightly over half of the noncertified growers, were likely to be grossing more than $5,000 per season from their farming operations. According to managers, market sales of organic farmers were comparable to or better than sales of other growers in the markets in this study.

Managers at several smaller markets in rural areas commented that some growers—full-time farmers as well as hobby gardeners—are developing an interest in organic production, even though they may be marketing in areas where customer demand is low or where experience with organic production among farmers and extension is limited. According to managers, this budding farmer interest is driven by personal, financial, and environmental motivations, as well as a desire to position themselves for a burgeoning customer base.

Most managers indicated that organic farmer participation has either increased or remained constant since the 2000 season, usually for two reasons. First, more farmers than in previous years were choosing to sell directly at farmers'markets while opting simultaneously to farm using organic methods. Second, the rapid growth in size and popularity of many markets was a drawing force for organic farmers, especially in situations where customer demand was compelling managers to search for additional organic growers.

Participation at the markets by certified organic growers had generally increased by one or two farmers within the previous 3 years, while participation by non-certified organic growers rose by one or more farmers yearly. Ten percent of markets studied either gained organic growers for the first time within the previous 3 years or had increased participation by organic growers. Markets with several organic growers often had a strong tendency to attract additional organic growers. Most markets without any participation by organic growers did not tend to gain any organic growers within the previous 3 years.

Managers indicated that customer interest in organic products at those markets was generally low.

Just over half of the managers indicated that the participation level of organic growers at their markets remained constant over the previous 3 years. In most cases, organic growers were reportedly successful at their markets and were likely to continue participating. At some markets where limited stall space was preventing the addition of other vendors, managers were planning to make room for organic growers to help meet strong customer demand. Some markets simply lacked organic farmers nearby who were not already committed to other markets.

In markets where the participation of organic growers was decreasing, managers explained that their organic farmers were moving on to bigger or wholesale markets, or were ceasing to farm for personal reasons not related to age. Managers often noted that organic farmers seemed younger than conventional farmers, and that they tended to drop out of markets for age-related reasons less frequently than conventional farmers.

Organic Standards and Certification

Congress passed the Organic Foods Production Act of 1990 to establish national standards for organically produced commodities, and USDA promulgated final rules that implemented this legislation on October 21, 2002. These regulations require that all organic growers and handlers (including food processors) be certified by a State or private agency accredited under the uniform standards developed by USDA, unless the farmers and handlers sell less than $5,000 a year in organic agricultural products. Retail food establishments that sell organically produced agricultural products but do not process them are also exempt from certification.

The national organic standards address the methods, practices, and substances used in producing and handling crops, livestock, and processed agricultural products. Although specific practices and materials used by organic operations may vary, the standards require every aspect of organic production and handling to comply with the provisions of the Organic Foods Production Act.

The labeling requirements under the national standards apply to raw, fresh products and processed foods that contain organic ingredients and are based on the percentage of organic ingredients in a product. Agricultural products labeled "100 percent organic" must contain (excluding water and salt) only organically produced ingredients. Products labeled "organic" must consist of at least 95 percent organically produced ingredients. Products labeled "made with organic ingredients" must contain at least 70 percent organic ingredients. Products with less than 70 percent organic ingredients cannot use the term organic anywhere on the principal display panel but may identify the specific ingredients that are organically produced on the ingredients statement on the information panel. In a processed product labeled as "organic," all agricultural ingredients must be organically produced unless the ingredient(s) is not commercially available in organic form. The USDA organic seal—the words "USDA organic" inside a circle—may be used on agricultural products that are "100 percent organic" or "organic." A civil penalty of up to $10,000 can be levied on any person who knowingly sells or labels as organic a product that is not produced and handled in accordance with these regulations.

Demand for Organic Products

Market managers reported that demand for organic products was strong in nearly 40 percent of the markets that had organic farmers, medium in 47 percent of these markets, and low in only 13 percent. Farmers' markets near major urban areas tended to have strong demand for organic products (map 2). Market communities near universities and other centers for higher education, near certain religious communities, or near facilities for holistic health care were also likely to exhibit strong demand. Markets with concentrations of organic growers, or one or two large-scale organic growers, tended to be associated with strong demand for organic products, regardless of setting. In rural areas, some markets had strong customer demand for organic products, according to managers, because they provided access locally to well-priced, fresh, organic foods otherwise only available in cities or towns some distance away. According to managers, customers with strong demand for organically grown products were most likely to exhibit interest in the social and environmental issues within agriculture, such as the relationship of agriculture to human and animal health, sustainable development, water scarcity, environmental pollution, and wildlife protection.

Growing interest in organic products led several managers to search for additional organic farmers to participate at their markets. Managers in California and Florida, in particular, indicated that there were not enough organic growers available to supply the quantity and variety of organic products needed to meet customer demand. Managers in many other States—including Colorado, Illinois, Maryland, Minnesota, North Carolina, New York, Tennessee, Utah, and Wisconsin—indicated that favorable opportunities for organic growers existed at their markets.

According to managers, customer interest in organic products has developed through customer contact with organic growers at farmers'markets, other direct marketing relationships, and influences such as increased volume and visibility of organic products in other market venues. Managers indicated that it was typical for organic farmers to exert a strong influence at their markets by bringing a wider range of specialty products than other market vendors. Organic farmers have historically played the role of the reliable, "cornerstone" vendors at some markets, which was also influential in generating demand for organic products, according to managers.

Managers noted that customers did not usually decide to buy from organic farmers simply because their products were organically grown. According to managers, organic farmers tended to excel at markets if they also satisfied customer desire for other attributes, such as consistently high product quality, access to specialty crop varieties, and excellent customer service. For example, one manager from Liberty, New York, indicated, "Partly because of the wide variety and sheer beauty of their stands, organic growers do well at my market." Managers in Illinois explained that "[conventional] farmers will typically bring peas, beans, cabbage, and squash, while organic growers will bring lettuce mixes, sprouts, spinach, cauliflower, and edamame (edible soybeans)" and "different things like parsnips and leeks."

Managers indicated that organic farmers are likely to converse with customers who ask how to prepare specialty food items, or inquire about organic farming techniques and associated costs of production. Such communication, they noted, can foster relationships between farmers and customers that may persist, especially if a customer is also satisfied with a farmer's product quality. One manager at an all-organic market in Vermont explained that "the most important thing is establishing a relationship between the customers and the growers. New growers take 3 to 4 weeks to gain solid interest from customers who are waiting to see if a grower is going to be reliable with product quality and attendance...That connection is more important than [a] label."

At some markets in this study, customer demand for organic products has had a strong influence on some farmers who recently decided to transition to organic farming or lower-input farming practices. For example, at a few markets in western Washington, several large-scale vegetable and fruit farmers who were switching from wholesaling to direct marketing at farmers'markets found that transitioning to organic production could enhance their marketing potential. "'Are you organic?'and 'Are you certified?'are the most frequently asked questions" at one market in Seattle, according to the manager. The influence of customer demand for organic products on growers' production methods was most prominent at markets in Washington and New Mexico.

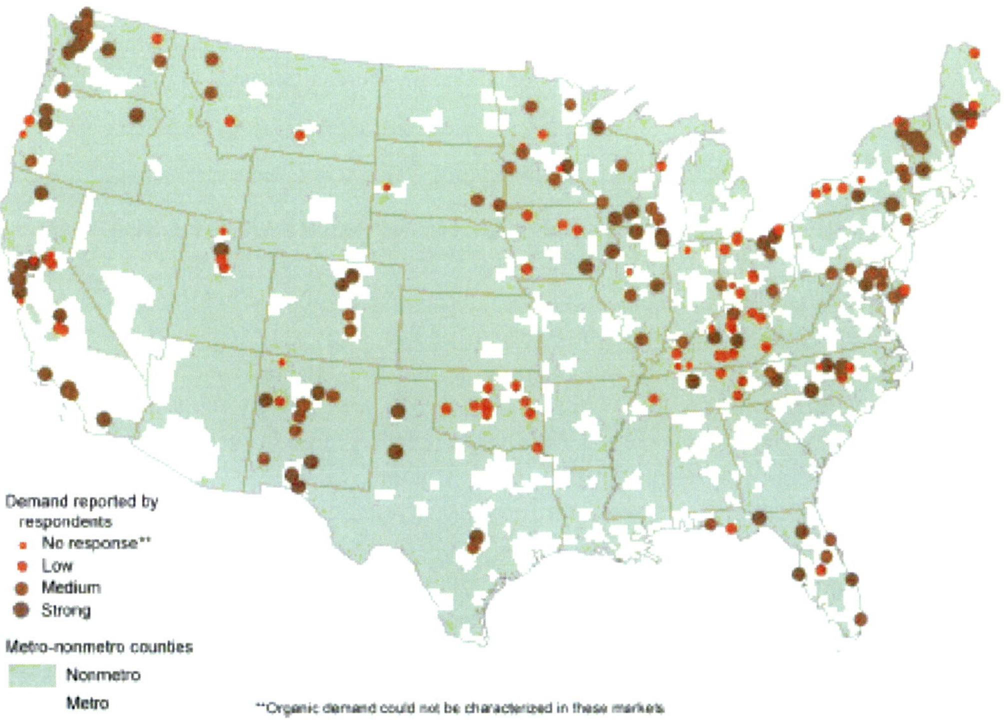

Map 2. Demand for organic products in 210 U.S. farmers' markets interviewed, 2002

Markets characterized in this study as having "medium" levels of demand for organic products had a small, often fast-growing, group of regular customers for organic products. Half of these markets were in towns or cities of less than 20,000 people, and had about half as many participating farmers as the average market in this study. A majority of the remaining markets with medium levels of demand were in cities of less than 100,000 people. Tourists visiting some markets in resort/vacation areas in Minnesota, New Mexico, and Washington contributed significantly to customer demand for organic products.

Markets where customers and farmers had little interest in—or negative opinions about—organically grown products were often in areas that managers described as being rural or still in early stages of organic food awareness. These markets had higher attendance by customers with limited incomes, compared with other markets. While customers had not developed much interest in organic products at these markets, several managers said that their customers had shown some interest in foods produced using fewer inputs. A few managers indicated that customers were reluctant to purchase organic products either due to previous negative experiences or expectations with regard to the quality or prices of organically grown products, or because organic growers had attended their markets too sporadically. Some market managers, particularly in the South, indicated that their farmers perceived organic farming as difficult, labor-intensive, and uneconomical due to pressures of weeds, pests, and diseases.

Education—about farming systems, sustainable agriculture, and how to prepare fresh food—was perceived by managers as the element most likely to generate additional interest in organic products and production from customers and conventional farmers. Several managers also suggested that relieving the short supply of organic products at their markets would likely stimulate interest and demand. For example, as one manager from New York explained, "People tend to think that everything at the market is organic, and when they find out that everything isn't, they are disappointed and want more. It's easier to make the market grow with more supply [of organic products]."

ECO-LABELS AND DIALOGUE EDUCATE MARKET CONSUMERS

Farmers'markets often receive coverage by the popular press, which helps offset small advertising budgets. Most managers interviewed for this study said they used very simple, low-cost techniques—such as articles, flyers, public service announcements for radio broadcast, websites, and notices for church and nonprofit newsletters—to attract farmers and customers to their markets. Managers generally promoted organic farmers and products only when organic farmers comprised over half of the vendors. Managers were reluctant to risk alienating their conventional growers with organic advertising or to inadvertently create a false impression that all products at their markets were organic. A small group of managers in nonmetro areas noted that advertising organic products and farmers would not be particularly effective, either due to negative perceptions or because the term organic was not "part of the community's vocabulary."

Markets with a strong history of organic farmer participation were more likely to advertise organic products and growers, as were those markets supported by nonprofit organizations. At some markets, where organic growers were a recent addition or where customer demand for organic products had grown significantly, managers were considering the use of some organic advertising for the 2003 season. Managers used an array of techniques—including color-coded market banners and price tags, detailed brochures, and website profiles of growers—to help orient and inform customers about certified organic, transitional, "pesticide-free," and other farming methods. Some techniques used by managers to promote their market, such as popular "Chef-at-Market" programs, often highlighted organic products and farmers.

Though some managers guide their farmers on how to successfully advertise their products, individual farm-ers'efforts vary. Some managers expressed surprise—or frustration—that organic farmers did not emphasize their organic attribute more, while others explained that only recently had shifts in customer awareness and interest in organic products made it more advantageous for farmers to advertise their organic growing practices. For example, one manager from Illinois indicated that her "customers feel like they can quit being afraid of organic, with more and more media being positive about it." Managers at a few markets, particularly in rural areas, indicated that organic product advertising was unlikely or still a few years away.

According to managers, talking was the main means of advertising used by non-certified organic growers to inform customers about products in 2002, while certified organic growers were more likely to use some kind of signage to promote their products. A range of signage types was used by growers, including individual labeling of products with the word "organic," displaying banners or plaques from certification agencies and farmer associations, and incorporating the word "organic" in farm names. Managers observed that some conventional producers have been using signs to indicate which crops, like lettuce and some fruit crops, had been grown without synthetic inputs.

Slightly under half of the organic growers in the markets studied during the 2002 market season were certified. Federal organic regulations requiring certification of products labeled organic were implemented after the 2002 market season had ended (see box, p. 5). Even prior to implementation of these regulations, about 15 percent of the markets studied did not allow noncertified organic produce to be labeled as organic. Other labels being used by organic farmers and others using alternative production methods were:

- "Chemical-free," "no chemical fertilizers," and "no harmful insecticides"
- "Natural growing conditions" and "naturally-grown"
- "Healthy farming practices used"
- "No-spray" • "Sustainably grown"
- "Authentic"
- "Free-range," "pastured," "grass fed," and "no antibiotics or hormones" • "Organically inclined"
- "Transitional" • "Wisconsin organic grown" and other local foods labels
- "Good bugs at work here"
- "Ask me how I grow this."

Organic farmers'advertising efforts also included picture displays of farms, recipes and business cards, websites, informational brochures, and farm tours.

Organic Price Premiums Differ across Markets

According to managers, prices for organic products were determined both casually and formally by individual farmers or by organic farmers as a group, sometimes with the help of pricing services such as AMS *Market News*, State price reports, and agricultural publications such as *Growing for Market*, which publish some information on wholesale prices for organic crops.

The use of price premiums for organic crops depended on seasonality and individual farmer choice. At about 40 percent of the markets where they participated, organic farmers did not request premium prices. Organic farmers charged a premium for their products, often modest, at about half of the markets where they participated. In some markets, there was no basis for comparison because the markets were all-organic or the organic and conventional growers sold different items. According to managers, organic farmers tended to set significant premiums only for products they perceived as being of exceptionally high quality, especially rare, or for markets in upscale communities. Premiums charged for poor quality product were unusual, and poorly received by farmers'market shoppers.

Organic farmers were more likely to ask for premiums when attending large markets where organic farmers represented half or more of the growers. Organic farmers participating at these markets generally had enough production to supply the market with a size-able volume of product, and were likely to be full- or part-time growers earning a significant portion of their household income from their farm operations.

Table 2. Advertising techniques used by organic growers in 210 U.S. farmers markets, 2002

Organic advertising techniques[1]	No. farmers
Talking (interaction)	187
Signs and labeling advertising organic methods	129
Posting of organic certification	
banners/plaques	100
Brochures, newsletters, business	
cards, and recipes	61
Photos	21

1 Some farmers use multiple techniques.

Certified and noncertified organic growers were equally likely to charge premiums for organic products at these markets.

Based on this information about the rarity and modesty of price premiums, some customers'reluctance to purchase organic foods may have less to do with actual prices than with expected prices for organic products. According to one manager in Maine, for example, "people have an image that organic is more expensive and may pass you by if you advertise too much."

Conclusion

Farmers'markets are providing an important marketing outlet for many farmers, including organic farmers, who want to generate substantial income from high-value crops grown on minimal acreage. Fair pricing, product quality and consumer trust are essential to the overall success of both the conventional and organic farmers participating in these markets. It is not unusual for consumers to visit farmers'markets thinking that most or all of the fresh produce sold there is organically produced. Farmers'markets provide a natural arena where organic farmers can educate consumers about what distinguishes their crops from the others being offered.

Information collected from farmers'market managers for this study suggest that demand for organic products is substantial and growing in many U.S. farmers'mar-kets. Demand for organic products and participation by organic growers at markets may be dampened by:

- Limited numbers of local organic farmers available to sell at markets,
- Limited awareness and interest of some consumers and farmers in organic production systems, and
- Negative perceptions of organic products or organic product pricing.

Many producers, even large-scale fruit and vegetable farmers, have become interested in incorporating organic production systems or direct marketing channels into their farming operations. Further investigation of direct marketing could shed light on the significance of net income derived from organic product sales at these markets, identify ways that organic growers acquire and use their direct marketing expertise, and clarify other poorly understood aspects of this marketing channel.

ACKNOWLEDGMENTS

The authors gratefully acknowledge the comments and reviews of Kitty Smith, Joy Harwood, and Keith Wiebe (ERS); John Love (WAOB); Barbara Robinson (AMS); and Gail Feenstra (University of California, Davis) and Garry Stephenson (Oregon State University, Corvallis). The authors also thank Chris Dicken for excellent geographic information system support and Dale Simms for valuable editorial assistance. We are also indebted to the farmers' market managers and State representatives who shared their observations and expertise with us for this project.

REFERENCES

Birkerts, Erik, G. (2000). "*The Local Market Opportunity for Organic Produce:An Objective Business Analysis,*" Prairie Partners Group, LLC.

Brown,Allison. (2002). "Farmers market research 1940-2000:An inventory and review." *American Journal of Alternative Agriculture*,Vol. *17*, No. *4*, 167-176.

Duram, Leslie, A. (1999). "Factors in organic farmers' decision making: Diversity, challenge, and obstacles," *American Journal of Alternative Agriculture*, Vol. *14*, No. *1*, 2-10.

Dimitri, Carolyn and Nessa Richman. (2000). *Organic Food Markets in Transition*, Henry A Wallace Center for Agricultural and Environmental Policy, Winrock International.

Federal Register. (2000). *National Organic Program; Final Rule*. Table 1- U.S. *Organic Product Sales*, 1990-1999, Vol. *65*, No. 246, Dec. 21.

Fernandez-Cornejo, Jorge, Catherine Greene, Renata Penn and Doris Newton. (1998). "Organic vegetable production in the U.S.: Certified growers and their practices," *American Journal of Alternative Agriculture*,Vol. *13*, No. 2.

Handy, Charles, R., Phil, R. Kaufman, Kristen Park, & Geoffrey, M. Green, (2000). "Evolving Marketing Channels Reveal Dynamic U.S. Produce Industry," *FoodReview*, Vol. *23*, Issue 2, U.S. *Department of Agriculture*, Economic Research Service, May-August.

Lockeretz,William (Ed.). (1987). *Sustaining Agriculture Near Cities*. Soil and *Water Conservation Society*, Ankeny, Iowa.

Lyson, T. A., Gillespie, G. W. Jr., & Hilchey, D. (1995). "Farmers' markets and the local community: Bridging the formal and informal economy," *American Journal of Alternative Agriculture*,Vol. *10*, No. *3*, 108-113.

Payne, Tim. (2002). *U.S. Farmers Markets-2000, A Study of Emerging Trends*, U.S. Department of Agriculture,Agricultural Marketing Service, *Transportation and Marketing Programs*, Marketing Services Branch.

Slama, Jim. (2002). *The Land of Organic Opportunity-Steps to Building a Regional Organic Food System Serving Chicago*, Sustain Alliance, Chicago, IL, www.sustainusa.org/local/organic/.

Stephenson, Garry, & Larry Lev. (1998). *Common Support for Local Agriculture in Two Contrasting Cities*, Oregon State University, *OSU Extension Service*, Oregon Small Farms Technical Report Number 1.

Walz, Erica. (1999). "*Third National Organic Farmers' Survey*," Organic Farming Research Foundation, Santa Cruz, California.

INDEX

A

B

C

F

G

H

I

J

K

L

M

N

O

P

T

U

V

W

Y